Multivalued Maps and Differential Inclusions
Elements of Theory and Applications

Multivalued Maps and Differential Inclusions

Elements of Theory and Applications

Valeri Obukhovskii

Voronezh State Pedagogical University, Russia

Boris Gel'man

Voronezh State University, Russia

 **World Scientific**

NEW JERSEY · LONDON · SINGAPORE · BEIJING · SHANGHAI · HONG KONG · TAIPEI · CHENNAI · TOKYO

Published by

World Scientific Publishing Co. Pte. Ltd.

5 Toh Tuck Link, Singapore 596224

USA office: 27 Warren Street, Suite 401-402, Hackensack, NJ 07601

UK office: 57 Shelton Street, Covent Garden, London WC2H 9HE

Library of Congress Control Number: 2020013498

British Library Cataloguing-in-Publication Data
A catalogue record for this book is available from the British Library.

MULTIVALUED MAPS AND DIFFERENTIAL INCLUSIONS
Elements of Theory and Applications

ISBN 978-981-122-021-0 (hardcover)
ISBN 978-981-122-022-7 (ebook for institutions)
ISBN 978-981-122-023-4 (ebook for individuals)

For any available supplementary material, please visit
https://www.worldscientific.com/worldscibooks/10.1142/11825#t=suppl

To Irina, Andrey, Alexey
and Barbara, Maxim and Sergey
with love...

Preface

The theory of multivalued maps and the theory of differential inclusions are closely related branches of contemporary mathematics which stood out as independent scientific disciplines around the middle of the last century. At the present time they continue to develop very rapidly, finding new supporters and new applications. A highly effective use of ideas and methods of the multivalued analysis and the theory of differential inclusions in such directions as theory of optimization and control, calculus of variations, nonsmooth and convex analysis, theory of differential equations, theory of games, mathematical economics and others, has become generally recognized and attracts the attention of many researchers all over the world.

In spite of several monographs and a huge number of other publications, in our opinion, the idea of a small book that provides a fairly basic introduction to the subject for "beginners", starting with students of senior courses and graduate students, is sufficiently relevant. We hope that the given book will be of interest both for theorists and persons who are involved in applied aspects of science.

Basing on the stuff included in this book, the authors were repeatedly lecturing for students of the Voronezh State University and the Voronezh State Pedagogical University.

The main exposition is preceded with the zero chapter, where necessary definitions and preliminary information, mainly from general topology are presented.

The first chapter of the book begins with examples showing how naturally the idea of a multivalued map arises in various branches of mathematics. Further, the types of continuity of multivalued maps as well as various operations over them and their properties are described. Then the concept of a continuous single-valued selection of a multivalued map is introduced and the classical Michael continuous selection theorem is proved. The concept of a single-valued approximation is discussed and the corresponding existence theorem is presented. We give analogues of above statements for multivalued maps with decomposable values. The first chapter concludes with the description of the properties of measurable multivalued functions. We give here the proof of the Filippov implicit function lemma which is known by its applications in control theory and study in detail the properties of the multivalued superposition operator which will be used while the discussing of differential inclusions. A statement on properties of a multivalued integral and the description of a superposition multioperator generated by an almost lower semicontinuous multimap are given.

The second chapter is devoted to the fixed point theory of multivalued maps. We produce here the Nadler theorem, a multivalued analogue of the classical Banach contraction map principle. We consider the properties of contraction multivalued maps depending on a parameter and describe the topological structure of the set of fixed points for a contractive multimap. Further, fixed point results are applied to the study of some classes of equations with surjective linear operators. We give also versions of the Caristi and Nemytski fixed point theorems. Further, we expose the topological degree theory for compact multivalued vector fields with convex values in a Banach space and present its applications to a number of fixed point results including the known Kakutani–Bohnenblust–Karlin theorem. Then the generalization of this theory to the case of condensing multivalued vector fields is described. The chapter is concluded with the study of topological properties of fixed point sets and the Browder–Ky Fan fixed point theorem and its application to the solving of variational inequalities.

The entire third chapter is devoted to the study of differential inclusions and their applications in control theory. We start with a series of examples, illustrating the appearance of differential inclusions in the description of control systems, in differential equations with discontinuous right-hand part and in mathematical economics. Then, based on the topological methods developed in the previous chapter, we give theorems on the existence of a solution to the Cauchy problem for differential inclusions of various types and differential equations with discontinuous right-hand part and describe the properties of solution sets. The topological degree theory finds the systematic application in the study of a periodic problem for differential inclusions. In particular, we expose in sufficient detail the method of guiding functions which at the present time is one of the most effective tools for the solving of periodic problems. Next we consider the question on the equivalence of control systems and differential inclusions and consider on that base applications to some optimization problems.

The last chapter is devoted to applications in the theory of dynamical systems, theory of games and mathematical economics. We describe main properties of generalized dynamical systems and their trajectories. This section incudes also the question on rest points of dynamical systems, which is solved by using the technique of the fixed point theory. Further, by applying a fixed point theorem for multivalued maps we prove the general equilibrium theorem in a two-person game which yields, as a consequences, the classical results on equilibrium in a zero-sum game and in a matrix game. It is shown also how the same fixed point methods provide the existence of an equilibrium in the Arrow–Debreu–McKenzie model of a competitive economy.

The book is concluded with the chapter named "Bibliographical comments and additions". We provide here fairly detailed comments that relate both to the sections described in the book and to those which left outside of its framework. For example, we give clues to the topological degree theory for multivalued maps with non-convex values, to the theory of differential inclusions in Banach spaces and to other actual parts of the theory.

We would like to express our deep gratitude to our colleagues and friends Aram Arutyunov, Anatoly Baskakov, Irene Benedetti, Yuri Gliklikh, Mikhail Kamenskii, Sergei Kornev, Yeong-Cheng Liou, Zhenhai Liu, Nguyen Van Loi, Luisa Malaguti, Paolo Nistri, Garik Petrosyan, Valentina Taddei, Ngai-Ching Wong, Jen-Chih Yao, Pietro Zecca for valuable and helpful discussions and support while working at this book. We are obliged to Alexei Gel'man and Irina Obukhovskaya for their help in preparation of the manuscript.

At last, we have the pleasure to express our thanks to the editors of the World Scientific Publishing for their constructive cooperation.

V. Obukhovskii, B. Gel'man
Voronezh, February 2020.

Contents

Chapter 0

Preliminaries

In order to know something you first need to know something.

—*Stanislaw Lem*

This chapter contains the preliminary information mainly from the general topology which is necessary for the further reading (details may be found, for example, in [1], [131], [135], [241], [270]). A reader familiar with these topics can pass on directly to Chapter 1.

We will use standard symbols $x \in X$ $(x \notin X)$, $X \subset Y$, $\complement X = Y \setminus X$ to denote *the belonging (not belonging) of an element to a set, the inclusion of a set into a set* (notice that the symbol $\subset$ does not exclude the equality of sets X and Y) and *the complement of a set X with respect to the whole space Y*. If $\{X_\alpha\}_{\alpha \in \mathcal{A}}$ is a certain family of sets, by the symbols $\bigcup\limits_{\alpha \in \mathcal{A}} X_\alpha$ and $\bigcap\limits_{\alpha \in \mathcal{A}} X_\alpha$ we denote *the union* and, respectively, *the intersection* of the sets of this family. *The Cartesian product* of sets X and Y is denoted by $X \times Y$. By the symbol $\{x | M(x)\}$, the set of objects x possessing a property $M(x)$ is denoted.

Let X be a set; by *a topology* on X we mean a system τ of subsets of X satisfying the following conditions:

1) $\emptyset$ and X belong to τ;
2) if U and V belong to τ then their intersection $U \cap V$ belongs to τ;
3) the union of each family of sets from τ belongs to τ.

Elements of the system τ are called *open sets*, and the pair (X, τ) is said to be *a topological space*. If the topology τ is implicitly meant we say simply about the topological space X.

If τ is a topology on a space X then its *base* is a subsystem $\tau_1 \subset \tau$ such that each element from τ can be represented as the union of a certain family of elements from τ_1.

The set of real numbers usually is endowed with the topology whose base consists of intervals (a, b).

Let (X, τ) be a topological space; *a neighborhood of a point* $x \in X$ is any subset of X in which an open set containing x lies. Similarly, any subset of X in which an open set containing a subset $A \subset X$ lies is called *a neighborhood of a set* A. A subset of X is open if and only if it is a neighborhood for each of its points.

The interior of a set A is the largest open set contained in A, it will be denoted as *int* A. The points of *int* A are called *interior points* of A.

A subset A of a topological space X is called *closed* if its complement $X \setminus A$ is open. *The closure* $\overline{A}$ of a set A is the least closed set containing A. A set is closed if and only if it coincides with its closure.

A topological space X is said to be *separable*, if it contains a countable subset A which is dense in X, i.e., $\overline{A} = X$.

Let $(X_1, \tau_1), (X_2, \tau_2)$ be topological spaces; the topology in the Cartesian product $X_1 \times X_2$ is generated in the following way: its base consists of the sets having the form $U_\alpha \times U_\beta$, where $U_\alpha \in \tau_1, U_\beta \in \tau_2$. The set $X_1 \times X_2$ endowed with such topology is called *the topological product* of X_1 and X_2.

A topological space X is called:

   (i)   *T_1-space* if each one-point set in X is closed;
   (ii) *Hausdorff* if any two different points from X possess disjoint neighborhoods;
   (iii) *regular* if it is a T_1-space such that any point from X and any closed subset to which it does not belong, possess disjoint neighborhoods;
   (iv) *normal* if it is a T_1-space such that any two its disjoint closed subsets possess disjoint neighborhoods.

Let X, Y be topological spaces; a map $f : X \to Y$ is *continuous* if the set $f^{-1}(V) = \{x | x \in X, f(x) \in V\}$ is open in X for each open set $V \subset Y$.

A topological space X defines on each of its subsets $A \subset X$ the topology whose open sets are the intersections of A with open subsets of X. Such topology is called *relative* or *induced*. The subset A with this topology is said to be *a subspace* (of a space X). If A is a subspace of X then a map $i : A \to X$ defined by the rule $i(x) = x$ is called *the inclusion map*. It is easy to see that the inclusion map is continuous.

A subset A of a topological space X is called *connected*, if it can not be represented as the union of two nonempty disjoint open (in the relative topology) sets. An open connected subset of a topological space is said to be *a domain*.

Let X be a topological space; a function $f : X \to \mathbb{R}$ is called *upper [lower] semicontinuous at a point* $x \in X$ if for every $\varepsilon > 0$ there exists a neighborhood $U(x)$ of the point x such that $f(x') < f(x) + \varepsilon$ for all $x' \in U(x)$ [respectively, $f(x') > f(x) - \varepsilon$ for all $x \in U(x)$]. If a function f is upper [or lower] semicontinuous at each point of a space X it is called *upper [or, respectively, lower] semicontinuous*. It is easy to see that a function f is upper [lower] semicontinuous if for every $r \in \mathbb{R}$ the set

$$\{x|x \in X, f(x) < r\},$$

respectively,

$$\{x|x \in X, f(x) > r\}$$

is open. While considering upper semicontinuous functions, it is often convenient to assume that they act into the extended set of real numbers $\bar{\mathbb{R}}$ obtained from $\mathbb{R}$ by addition of $+\infty$ and $-\infty$.

Let $\mathcal{A}$ be a set with a given binary relation $\leq$. The set $\mathcal{A}$ is called *directed* if the following conditions hold:

1) $\alpha \leq \beta$, $\beta \leq \gamma$ imply $\alpha \leq \gamma$ for every $\alpha, \beta, \gamma \in \mathcal{A}$;
2) $\alpha \leq \alpha$ for every $\alpha \in \mathcal{A}$;
3) for every $\alpha, \beta \in \mathcal{A}$ there exists $\gamma \in \mathcal{A}$ such that $\alpha \leq \gamma$, $\beta \leq \gamma$.

A map of a directed set $\mathcal{A}$ into a topological space X, i.e., the correspondence which assigns to each $\alpha \in \mathcal{A}$ a certain $x_\alpha \in X$ is said to be *a net* or *a generalized sequence*.

A net $\{x_\alpha\} \subset X$ *converges to a point* $x \in X$ if for every neighborhood U of the point x there exists an index α_0 such that $x_\alpha \in U$ for all $\alpha \geq \alpha_0$. A point x belongs to the closure of a subset M of a space X if and only if M contains a net converging to x. In the case of a metric space X a net here may be substituted with a usual sequence.

A *cover of a set* X is a collection Σ of subsets of X whose union is the whole X. A cover $\sum'$ is called *subcover* of a cover Σ provided each of sets from the collection $\sum'$ belongs to Σ. If each cover of a topological space X by open sets contains a finite subcover then the space X is called *compact*.

The compactness of a space X is equivalent to each of the following conditions:

1) each net in X contains a convergent subnet;

2) each centered collection of closed subsets of X (i.e., such collection that each its nonempty finite subcollection has a nonempty
intersection) also has a nonempty intersection.

A set X is said to be *relatively compact*, if its closure $\overline{X}$ is compact. A
upper semicontinuous function $f : X \to \mathbb{R}$ defined on a compact space X
reaches its maximum, whereas a lower semicontinuous function reaches its
minimum.

By virtue of the Tychonoff theorem the topological product $X_1 \times X_2$ of
compact spaces X_1 and X_2 is compact.

A subset of the Euclidean n-dimensional space $\mathbb{R}^n$ is relatively compact
if and only if it is bounded.

A cover Σ of a topological space X is called *locally finite* if every point
$x \in X$ possesses a neighborhood U which intersects only a finite number
of sets from Σ. A topological space X is said to be *paracompact* if it is
Hausdorff and each its open cover $\triangle$ has an open locally finite refinement
Σ (i.e., each of the sets from Σ is contained in a set from $\triangle$).

For each locally finite open cover $\Xi = \{U_j\}_{j \in J}$ of a paracompact space
X there exists a subordinated *partition of unity*, i.e., a family $\{p_j\}_{j \in J}$ of
continuous on X nonnegative functions such that:

1) for each $j \in J$ we have: $supp\, p_j = \{x | x \in X, p_j(x) \neq 0\} \subset U_j$;
2) for each point $x \in X$: $\sum_{j \in J} p_j(x) = 1$.

Notice that due to the local finiteness of the cover Ξ, only a finite number
of terms in the last sum differ from zero.

Let (X, ϱ) be a metric space, $x \in X$, and $r > 0$. The set

$$B_r(x) = \{y | y \in X, \varrho(x, y) < r\}$$

is called *an open ball* of the radius r with the center at x, whereas the set

$$\overline{B}_r(x) = \{y | y \in X, \varrho(x, y) \leq r\}$$

is *a closed ball* of the radius r with the center at x. The collection of all
open balls is the base of a certain topology on X which is called *metric
topology*. It is clear that a set V in a metric space X is open if and only if
every point x of V belongs to V with a certain open ball centered at x.

Two metrics on a set X are called *equivalent* if they generate on X
the same metric topology. Each space with metric topology is normal and
hence regular and Hausdorff. Due to the Stone theorem every metric space
is paracompact.

The distance from a point x to a set $A \subset X$ is defined as

$$\varrho(x, A) = \inf \{\varrho(x, y) \,|\, y \in A\}.$$

If $A \subset X$ and $\varepsilon > 0$ then the set

$$U_\varepsilon(A) = \{y | y \in X, \varrho(y, A) < \varepsilon\}$$

is called an ε-neighborhood of the set A.

Let A be a compact metric space and Σ an open cover of X. Then, according to the Lebesgue covering lemma (see, e.g., [241]), there exists a positive real number r with the property that for each $x \in X$ there exists a set $U \in \Sigma$ such that $B_r(x) \subset U$. From this assertion it follows that if (X, ϱ) is a metric space, A is a compact subset of X, B is a closed subset of X and $A \cap B = \emptyset$ then

$$\varrho(A, B) = \inf \{\varrho(a, b) \,|\, a \in A, b \in B\} > 0$$

and hence there exists such $\varepsilon > 0$ that

$$U_\epsilon(A) \cap B = \emptyset.$$

This yields, in particular, that for every open neighborhood U of a compact set A there exists a sufficiently small $\varepsilon > 0$ such that the ε-neighborhood of A is contained in U.

Let T be a compact space, (X, ϱ) a metric space. On the set $C(T, X)$ of all continuous functions from T to X the matric $\tilde{\varrho}$ may be defined by the formula

$$\tilde{\varrho}(f_0, f_1) = \sup_{t \in T} \varrho(f_0(t), f_1(t)).$$

The topology τ_c generated on $C(T, X)$ by this metric is called *the topology of uniform convergence*.

Let (T, ϱ_T) be a compact metric space. A family of functions $H \subset C(T, X)$ is called *equicontinuous* if for each $\varepsilon > 0$ there exists $\delta > 0$ such that for every $t, t' \in T$ the condition $\varrho_T(t, t') < \delta$ implies $\varrho(f(t), f(t')) < \varepsilon$ for all $f \in H$. According to the Arzela–Ascoli theorem (see, e.g., [241]) *if a subset $H \subset C(T, X)$ is equicontinuous and the sets $H(t) = \{f(t) | f \in H\}$ are relatively compact in X for all $t \in T$ then H is relatively compact in the space $(C(T, X), \tau_c)$.*

If X is a linear space and $A, B \subset X$ then

$$A + B = \{a + b \mid a \in A, b \in B\}.$$

If $\alpha \in \mathbb{R}$ then

$$\alpha A = \{\alpha a \mid a \in A\}.$$

The set of all finite linear combinations

$$\sum_i \lambda_i x_i,$$

where $\lambda_i \geq 0, \sum_i \lambda_i = 1$ and every x_i belongs to A is a least convex set containing A and it is called *the convex hull of the set A* and is denoted as coA.

Let X be a linear space on which a topology τ be defined. The pair (X, τ) is said to be *a linear topological space* if the topology τ is consistent with the linear operations on X in the following way: 1) the addition operation is continuous, i.e., the map

$$X \times X \to X, (x, y) \to x + y;$$

is continuous; 2) the number multiplication operation is continuous, i.e., the map

$$X \times \mathbb{R} \to X, (x, \lambda) \to \lambda x$$

is continuous.

If X is a linear topological space and $A \subset X$ then the closure of the set coA is denoted by $\overline{co}A$. It is called *the convex closure A* and it is a least convex closed set containing A.

The following Brouwer fixed point theorem holds true. *If M is a convex closed subset of a finite-dimensional linear topological space then every continuous map $f : M \to M$ such that its range $f(M)$ is bounded has at least one fixed point $x \in M$, $x = f(x)$.*

We will suppose that the reader is familiar with the concepts of normed and Banach spaces as well as with a main information concerning their properties (see, e.g., [120], [124], [247], [371], [384] and others).

Nevertheless, let us indicate the following facts that we will use in the sequel.

Let A be a closed subset of a metric space X and Y a normed space. Then each continuous map $f : A \to Y$ has a continuous extension $\widetilde{f} : X \to Y$ and, moreover $\widetilde{f}(X) \subset cof(A)$ (the Tietze–Dugundji theorem). The following assertion is an immediate consequence of this result. *If Y is a normed space and A is its nonempty closed convex subset then there exists a continuous map (the retraction) $r : Y \to A$ such that $r(y) = y$ for all $y \in A$.*

If X is a Banach space and $A \subset X$ is a compact set then its convex closure $\overline{co}A$ is also compact (the Mazur theorem).

If A is a bounded subset of a normed space X then by *the norm of the set A* we mean the value

$$\|A\| = \sup_{a \in A} \|a\|.$$

We will assume also that the reader is familiar with the notions of *the Lebesgue measure,* a *measurable* and a *Bochner integrable* function with the values in a Banach space as well as with main properties of the space of integrable functions L^1(see, e.g., [124], [316], [371], [384], [408]).

The sign := will denote the equality by definition.

The end of the proof will be marked with the symbol ■.

Chapter 1

Multivalued maps

And I claim that it is sufficient to launch any fulcrum into space and to place a ladder to it. The road to heaven is open!

—Stanislaw Jerzy Lec

1.1 Some examples

Mathematics is the part of physics in which the experiments are very cheap.

—Vladimir Arnold

Let X and Y be arbitrary sets; *a multivalued map (multimap)* F of a set X into a set Y is the correspondence which associates to every $x \in X$ a nonempty subset $F(x) \subset Y$, called *the value* (or *the image*) of x. Denoting by $P(Y)$ the collection of all nonempty subsets of Y we can write this correspondence as

$$F : X \to P(Y).$$

It is clear that the class of multivalued maps includes into itself usual single-valued maps: for them each value consists of a single point.

In the sequel we will denote multimaps by capital letters.

Definition 1.1.1. For any set $A \subset X$ the set $F(A) = \bigcup_{\alpha \in A} F(\alpha)$ is called *the image of the set A under the multimap F.*

Definition 1.1.2. Let $F : X \to P(Y)$ be a multimap. The set Γ_F in the Cartesian product $X \times Y$,

$$\Gamma_F = \{(x, y) \mid (x, y) \in X \times Y, \ y \in F(x)\}$$

is called *the graph* of the multimap F.

It is worth noting that the concept of a multimap is not something too unusual: after all, we encounter with maps of this kind already in elementary mathematics when trying to invert, for example, such functions as $y = x^2$ or $y = \sin x$ and others. However, here the "non-singlevaluedness" of the inverse function is perceived, rather, as a negative circumstance: the introduction of such notions as arithmetic value of the square root, or functions of type arcsin, arccos etc. is related precisely with the "liquidation" of this ambiguity.

Consider a few examples of multimaps.

Example 1.1.3. Denote pr_1, pr_2 the projections from $X \times Y$ onto X and Y respectively. Each subset $\Gamma \subset X \times Y$ such that $pr_1(\Gamma) = X$ defines the multimap $F : X \to P(Y)$ by the formula $F(x) = pr_2 \circ pr_1^{-1}(x)$. It is clear that the graph Γ_F of the multimap F coincides with Γ.

Example 1.1.4. Define the multimaps of the interval $[0,1]$ into itself assuming

(a) $F_1(x) = [x, 1]$;

(b) $F_2(x) = \begin{cases} [0, 1/2], & x \neq 1/2, \\ [0,1], & x = 1/2; \end{cases}$

(c) $F_3(x) = \begin{cases} [0,1], & x \neq 1/2, \\ [0, 1/2], & x = 1/2. \end{cases}$

The graphs of these multimaps are presented in Fig. 1–3.

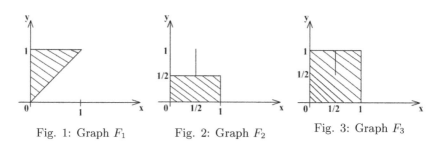

Fig. 1: Graph F_1 Fig. 2: Graph F_2 Fig. 3: Graph F_3

Denote $\overline{\mathbb{R}}_+ = \{x | x \in \mathbb{R}, x \geq 0\}$.

Example 1.1.5. Define the multimap $F : [0, \frac{\pi}{2}) \to P(\overline{\mathbb{R}}_+)$, assuming $F(x) = [\tan x, +\infty)$ (Fig. 4).

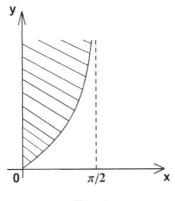

Fig. 4

Example 1.1.6. Define the multimap $F : [0, \pi] \to P(\mathbb{R})$,

$$F(x) = \begin{cases} [\tan x, 1 + \tan x], & x \neq \frac{\pi}{2}, \\ \{0\}, & x = \frac{\pi}{2}. \end{cases}$$

The graph of the multimap F is shown in the Fig. 5.

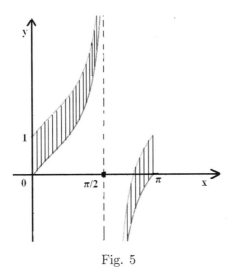

Fig. 5

Example 1.1.7. Define the multimap $F : \overline{\mathbb{R}_+} \to P((0, 1])$ assuming $F(x) = [e^{-x}, 1]$ (Fig. 6):

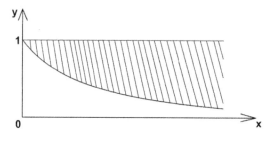

Fig. 6

Example 1.1.8. Define the multimap $F : \mathbb{R}^2 \to P\left(\mathbb{R}^2\right)$ assuming for $x = (x_1, x_2) \in \mathbb{R}^2$:

$$F(x) = \{(y_1, y_2) \mid (y_1, y_2) \in R^2, (y_1, y_2) = (x_1 + z_1, x_2 + z_2),$$

$$z_1 > 0, z_2 > 0, z_1 \cdot z_2 = 1\}.$$

The multimap F (but not the graph Γ_F!) is shown in the Fig. 7.

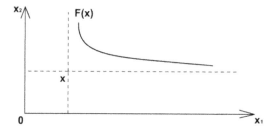

Fig. 7

Example 1.1.9 (Inverse functions). If X, Y are arbitrary sets and $f : X \to Y$ is a surjective map then the multimap $F : Y \to P(X)$, $F(y) = \{x \mid x \in X, f(x) = y\}$ is the inverse to f.

Example 1.1.10 (Implicit functions). Let X, Y, Z be arbitrary sets, maps $f : X \times Y \to Z$ and $g : X \to Z$ are such that for every $x \in X$ there exists $y \in Y$ such that $f(x, y) = g(x)$. The implicit function defined by f and g, in a general case, is the multimap $F : X \to P(Y)$, $F(x) = \{y \mid y \in Y, f(x, y) = g(x)\}$.

Example 1.1.11. Let X, Y be arbitrary sets, $f : X \times Y \to \mathbb{R}$ a function. Let for a certain number $r \in \mathbb{R}$ for every $x \in X$ there exists $y \in Y$ such that $f(x,y) \le r$. Then the following multimap $F_r : X \to P(Y)$ can be defined: $F_r(x) = \{y|y \in Y, f(x,y) \le r\}$.

Example 1.1.12. Generalized dynamical systems.

a) A multivalued translation operator.

Let a set X be the space of states of a certain dynamical system such that being at the initial moment in the state $x \in X$ this system may move further along various trajectories. For example, such situation holds if the behavior of the system is governed by a differential equation which does not satisfy the uniqueness of a solution condition or contains a control parameter. A generalized dynamical system is defined if its *reachable sets* $Q(x,t) \subset X$ are given, i.e., the sets of all states into which system can shift in the time $t \ge 0$ from the state $x \in X$ are indicated. The multimap $Q : X \times \mathbb{R}_+ \to P(X)$ arising in such a manner is called *the translation multioperator along the trajectories of the system*. Notice that usually the translation multioperator satisfies the natural conditions:

1) $Q(x,0) = \{x\}$;
2) $Q(x,t_1 + t_2) = Q(Q(x,t_1),t_2)$ for all $x \in X; t_1, t_2 \in \overline{\mathbb{R}}_+$.

b) Multivalued fields of directions.

Consider an important way of setting of a generalized dynamical system. Let $\mathbb{R}^n$ be the state space of a system and for every state $x \in R^n$ the set $F(x) \subset R^n$ of velocities with which the system can leave x be given. The multimap $F : \mathbb{R}^n \to P(\mathbb{R}^n)$ which is defined in such a manner is called *the multivalued field (multifield) of directions*. A function $x : \Delta \to \mathbb{R}^n$, where $\Delta \subset \mathbb{R}$ is a certain interval is called *an integral curve* of the multifield F if at every (or almost every) point $t \in \Delta$ it has the derivative $x'(t)$ and

$$x'(t) \in F(x(t))$$

for all (or almost all) $t \in \Delta$. Such a relation is called *a differential inclusion* and the integral curve x is its *solution*.

A sulution $x : \Delta \to \mathbb{R}^n$ is *the trajectory* of a given multifield of velocities. The collection $Q(x,t)$ of points of such trajectories at the moment t emanating from a given point $x \in \mathbb{R}^n$ defines the translation multioperator Q along the trajectories of the multifield F.

Suppose, for example, that considered generalized dynamical system is a control system whose dynamics is governed by a differential equation

$$x'(t) = f(x(t), u(t)),$$

where $f : \mathbb{R}^n \times \mathbb{R}^m \to \mathbb{R}^n$ is a map, $u(t) \in \mathbb{R}^m$ a control parameter. The feedback in this system is described by a multimap $U : \mathbb{R}^n \to P(\mathbb{R}^m)$ which defines for every given state $x \in \mathbb{R}^n$ a set of admissible controls $U(x)$. Then the multifield of directions for a given system is defined by the formula

$$F(x) = f(x, U(x)).$$

We will study differential inclusions, control systems and generalized dynamical systems in Chapters 3 and 4 in more detail.

Example 1.1.13. Metric projection. The following notion arises naturally in the theory of best approximations. Let (X, ϱ) be a metric space; $C \subset X$ a nonempty closed subset. For $x \in X$, the set $\mathbb{P}_C(x)$ of points $y \in C$ such that $\varrho(x, y) = \varrho(x, C)$ is called *the metric projection* of x onto C. Notice that the set $\mathbb{P}_C(x)$ may be empty. If $\mathbb{P}_C(x) \neq \emptyset$ for every $x \in X$ the set C is called *proximinal*. In this case there arises the multimap $\mathbb{P}_C : X \to P(C)$ which is also called the metric projection. As examples of proximinal sets may be considered compact sets as well as closed convex subsets of reflexive Banach spaces. Metric projections play an important role in various problems of the approximation theory, geometry of Banach spaces, fixed point theory, variational methods. Extensive literature is devoted to the study of their properties (see, e.g., [175], [380]).

Example 1.1.14. Approximate calculations. Suppose that at each point x of a certain set X some number characteristics $y(x) = (y_1(x), ..., y_n(x)) \in \mathbb{R}^n$ are measured. By the nonhomogeneity of the set X, absolute errors of measurements δ_i $(1 \leq i \leq n)$ depend on $x : \delta_i = \delta_i(x)$. The multimap $F : X \to P(\mathbb{R}^n)$,

$$F(x) = \{y | y \in \mathbb{R}^n, |y_i - y_i(x)| \leq \delta_i(x), 1 \leq i \leq n\},$$

is called *the field of values of characteristics*.

Example 1.1.15. Theory of games.

(a) **Zero-sum games.**

Notice that first examples and applications of the notion of a multi-valued map were connected with the new science, arising in the thirties-forties of the XX century, the *theory of games*. This branch of mathematics studies the mathematical models of conflict situations, i.e., such collisions in which the interests of participants do not coincide or are directly opposite. Situations of such kind emerge repeatedly in economics, military or political conflicts and in other spheres of human activity. Their simple and visual models are provided by chess, card games etc., from where the name of the discipline comes from. The base of the theory of games was laid by such prominent scientists as J. von Neumann, J.Nash, O.Morgenstern and others.

From the mathematical point of view the behavior of the participants of a conflict situation (let us call them *players*) is determined by the choice of *a strategies*, points from a certain sets of admissible strategies. The selection of a strategy completely defines the behavior of a player at each position which can arise in the process of a game. It is easy to see that even in very simple games there is an enormous number of possible strategies and so their analysis is not a very simple matter. What can be the main principles of such analysis?

For simplicity, let us consider the case of a game with two players, or *the two-person game*. Let all admissible strategies of the first player form a set X and Y be a set of all admissible strategies of the second player. By *a game rule* of the first player we mean the assignment to each strategy $y \in Y$ of the second player the set of best strategies $A(y) \subset X$ from which the first player chooses his strategy. Similarly, the game rule for the second player is defined by the sets of his strongest responses $B(x) \subset Y$ to the strategies $x \in X$ of the first player. This means that the game rule of the first player may be interpreted as the multimap $A : Y \to P(X)$, whereas the game rule of the second player is the multimap $B : X \to P(Y)$.

For a simple example of constructing of game rules we can consider *a zero-sum* or *antagonistic* game. The game of this type is determined by *the payoff function* $f : X \times Y \to \mathbb{R}$ defined on the Cartesian product of the spaces of strategies. It is supposed that after the choice by the first player of his strategy $x \in X$ and by the second player of the strategy $y \in Y$, the payoff of the first player is equal to $f(x, y)$ whereas the payoff of the second player is directly opposite and equals $-f(x, y)$. This means that the first player is trying to maximize the value $f(x, y)$ whereas the second one is making efforts to minimize it. In this case the game rules can be given explicitly:

$$A\left(x\right) = \left\{ y | y \in Y, \ f\left(x, y\right) = \min_{\tilde{y} \in Y} f\left(x, \tilde{y}\right) \right\}$$

$$B\left(y\right) = \left\{ x | x \in X, \ f\left(x, y\right) = \max_{\tilde{x} \in X} f\left(\tilde{x}, y\right) \right\}$$

of course, under condition that pointed out maximums and minimums exist.

Therefore, while the elaboration of suitable strategies, each of the players should analyze the multimaps $A : X \to P\left(Y\right)$ and $B : Y \to P\left(X\right)$. The consideration of the question how these multimaps may be used for the searching of optimal strategies for each player is postponed till the fourth chapter.

(b) **Games with a complete information.**

The language of multivalued maps allows also to simulate some game situations in the following way. Let X be a set of game positions partitioned into n subsets $X_1, ..., X_n$ in accordance with the number of players. For each player, a certain preference relation is given on X which allows him to compare positions from the point of view of their utility. Let $\{a\}$ be any singleton, $a \notin X$ and $F : X \to P\left(X \cup a\right)$ a multimap such that $a \in F\left(x\right)$ implies $F\left(x\right) = \{a\}$. Let an initial position $x_0 \in X_i$ be given, then the i-th player makes his move choosing a position x_1 in the set $F\left(x_0\right)$. If $x_1 \in X_j$ then the j-th player chooses a position in the set $F\left(x_1\right)$ and so on. The game is over if any player chooses a position x such that $F\left(x\right) = \{a\}$. The goal of the game for an individual player may be formulated as, for example, the obtaining of a position being as profitable as possible at least once during the game.

Example 1.1.16. Mathematical economics.

(a) **Multifunctions of productivity and demand.**

Let an economic system include n categories of goods whose prices $p = (p_1, ..., p_n)$ can vary in frameworks of a set $\Delta \subset \mathbb{R}^n$. Let an enterprise–producer has a certain compact set $Y \subset \mathbb{R}^n$ of possible production plans for the output of goods (a technological set). The component y_j of the vector $y \in Y$ corresponds to the amount of the j-th commodity produced

in accordance with this plan. The profit of the producer after the realiza-
tion of the plan y equals $\langle p, y \rangle = \sum\limits_{j=1}^{n} p_j y_j$. Be guided by considerations to
obtain the maximal gain under the prices $p \in \Delta$, the producer will choose
production plans from the set

$$\Psi(p) = \left\{ y | y \in Y, \ \langle p, y \rangle = \max_{\tilde{y} \in Y} <p, \tilde{y}> \right\}.$$

The multimap $\Psi : \Delta \to P(\mathbb{R}^n)$ defined in such a way is called *the productive
multifunction* of the enterprise.

On the other hand, let at given prices $p \in \Delta$ for the enerprise-customer
a compact set $X(p) \subset \mathbb{R}^n$ of consumption vectors be accessible. The component x_j of the vector $x \in X(p)$ corresponds to the consumption of the
j-th product. The preference of one or other consumption vectors is characterized by a certain function $u : \mathbb{R}^n \to \mathbb{R}$ which is called *the utility index*.
Trying to purchase at the given prices the most useful collection of goods,
the customer will make his choice in the set

$$\Phi(p) = \left\{ x | x \in X(p), \ u(x) = \max_{\tilde{x} \in X(p)} u(\tilde{x}) \right\}.$$

The multimap $\Phi : \Delta \to P(\mathbb{R}^n)$ is called *the demand multifunction* of the
customer.

A more detailed description of an economic model of that type and an
application of the multimaps techniques to the finding of an equilibrium in
it will be carried out in the fourth chapter.

(b) **Economic dynamics.**

Suppose that in an economic system a vector $x_{(t)} \in \mathbb{R}^n$ characterizes
the collection of goods produced by the moment t during the preceding unit
time interval (for example, a year). A part of this collection, $y_{(t)}$ comes to
consumption, whereas the remaining part $z_{(t)} = x_{(t)} - y_{(t)}$ is spend to the
accumulation, i.e., serves as the resource for the obtaining a new outcome
vector $x_{(t+1)}$. The pair $(y_{(t)}, z_{(t)})$ is called *the state of economics* at the
moment t. By investing the resource $z_{(t)}$ into accumulation, it is possible to
produce by the moment $t + 1$ one of collections of goods in the frameworks
of a certain set $B_t(z_{(t)}) \subset \mathbb{R}^n$. The multimap $B_t : \mathbb{R}^n \to P(\mathbb{R}^n)$ called
productive characterizes the technology of the system at the moment t. So,
starting from the state of economics $(y_{(t)}, z_{(t)})$ it is possible to obtain by
the next moment one of states filling the set

$$A_t(y_{(t)}, z_{(t)}) = \{(y_{(t+1)}, z_{(t+1)}) \mid (y_{(t+1)}, z_{(t+1)}) \in \mathbb{R}^n \times \mathbb{R}^n,$$

$$y_{(t+1)} + z_{(t++1)} \in B_t \left(z_{(t)} \right) \}.$$

Multimaps $A_t : \mathbb{R}^n \times \mathbb{R}^n \to P(\mathbb{R}^n \times \mathbb{R}^n)$ play an important role in the study of models of mathematical economics.

Example 1.1.17. Non-smooth optimization.

In contemporary optimization theory it is necessary very often to find the maximums and minimums of functions which are not differentiable. Functions of that kind arise, for example, while the transfer to suprema and infima of families of smooth functions. (So "classical" non-differentiable at zero function $y = | \, x \, |$ can be obtained as a supremum of functions $y = x$ and $y = -x$). For the searching of extrema of such functions, the notion of a derivative must be extended.

Let, for example, E be a finite-dimensional linear space; $f : E \to \mathbb{R}$ a convex functional. For a given $x \in E$ the set $\partial f(x) \subset E$ of all points $\xi \in E$ such that for all $v \in E$ we have

$$f(x+v) - f(x) \geq \ <\xi, v>$$

is called *the subdifferential* of a functional f at x.

So, for a given functional instead of an ordinary derivative we have to deal with a modified derivative, expressed by the multimap $x \to \partial f(x)$. The classical Fermat rule in this situation takes the following form: *if x_0 is a point of a local extremum of a functional f then $0 \in \partial f(x_0)$.*

It is easy to see that for the function $y = | \, x \, |$ the subdifferential is evaluated by the formula:

$$\partial f(x) = \begin{cases} -1, & x < 0; \\ [-1,1], & x = 0; \\ 1, & x > 0. \end{cases}$$

Concerning the problems of non-smooth analysis and methods for their solving see the monographs [24], [26], [27], [62], [104], [117], [118], [119], [120], [129], [216], [310], [311], [367] and others.

1.2　Continuity of multivalued maps

> He who wants to get to the source must swim against the current.
>
> *—Stanislaw Jerzy Lec*

The classical concept of continuity of a single-valued map splits into different notions when generalized to multimaps and each of these types of continuity has its own specific properties. This variety is based on the fact that the usual set-theoretic notion of the inverse image of a set can be interpreted differently when being applied to multimaps. We will start with the study of this notion.

1.2.1 *Small and complete preimages of a set*

Let X, Y be sets, $F : X \to P(Y)$ a multimap.

Definition 1.2.1. *The small preimage* of a set $D \subset Y$ is the set

$$F_+^{-1}(D) = \{x | x \in X, \ F(x) \subset D\}.$$

Definition 1.2.2. *The complete preimage* of a set $D \subset Y$ is the set

$$F_-^{-1}(D) = \{x | x \in X, \ F(x) \cap D \neq \emptyset\}.$$

It is clear that $F_+^{-1}(D) \subset F_-^{-1}(D)$.

Let $A \subset X; \ D \subset Y; \ \{D_j\}_{j \in J}$ a family of subsets of Y, J a set of indices.

The next properties of small and complete preimages follow immediately from the definitions (**verify!**).

Lemma 1.2.3.

(a) $F_+^{-1}(F(A)) \supset A$;

(b) $F\left(F_+^{-1}(D)\right) \subset D$;

(c) $\complement F_+^{-1}(D) = F_-^{-1}(\complement D)$;

(d) $F_+^{-1}\left(\bigcup_{j \in J} D_j\right) \supset \bigcup_{j \in J} F_+^{-1}(D_j)$;

(e) $F_+^{-1}\left(\bigcap_{j \in J} D_j\right) = \bigcap_{j \in J} F_+^{-1}(D_j)$.

Lemma 1.2.4.

(a) $F_-^{-1}(F(A)) \supset A$;

(b) $F\left(F_-^{-1}(D)\right) \supset D \cap F(X)$;

(c) $\complement F_-^{-1}(D) = F_+^{-1}(\complement D)$;

(d) $F_-^{-1}\left(\bigcup_{j \in J} D_j\right) = \bigcup_{j \in J} F_-^{-1}(D_j)$;

(e) $F_-^{-1}\left(\bigcap_{j \in J} D_j\right) \subset \bigcap_{j \in J} F_-^{-1}(D_j)$.

Let us observe the properties of small and complete preimages while passing to various set-theoretic operations on multimaps.

Definition 1.2.5. Let $F_0, F_1 : X \to P(Y)$ be multimaps. The multimap $F_0 \cup F_1 : X \to P(Y)$,

$$(F_0 \cup F_1)(x) = F_0(x) \cup F_1(x)$$

is called *the union of the multimaps F_0 and F_1.*

Definition 1.2.6. Let $F_0, F_1 : X \to P(Y)$ be multimaps such that $F_0(x) \cap F_1(x) \neq \emptyset$ for all $x \in X$. The multimap $F_0 \cap F_1 : X \to P(Y)$,

$$(F_0 \cap F_1)(x) = F_0(x) \cap F_1(x),$$

is called *the intersection of the multimaps F_0 and F_1.*

The following properties can be easily verified (**do it!**).

Lemma 1.2.7. *If $D \subset Y$ then*

(a) $(F_0 \cup F_1)_+^{-1}(D) = (F_0)_+^{-1}(D) \cap (F_1)_+^{-1}(D)$;

(b) $(F_0 \cap F_1)_+^{-1}(D) \supset (F_0)_+^{-1}(D) \cup (F_1)_+^{-1}(D)$.

Lemma 1.2.8. *If $D \subset Y$ then*

(a) $(F_0 \cup F_1)_-^{-1}(D) = (F_0)_-^{-1}(D) \cup (F_1)_-^{-1}(D)$;

(b) $(F_0 \cap F_1)_-^{-1}(D) \subset (F_0)_-^{-1}(D) \cap (F_1)_-^{-1}(D)$.

Definition 1.2.9. Let X, Y, and Z be sets, $F_0 : X \to P(Y)$, $F_1 : Y \to P(Z)$ multimaps. The multimap $F_1 \circ F_0 : X \to P(Z)$,

$$(F_1 \circ F_0)(x) = F_1(F_0(x))$$

is called *the composition of the multimaps F_0 and F_1.*

Lemma 1.2.10. *Let $D \subset Z$ then*

(a) $(F_1 \circ F_0)_+^{-1}(D) = (F_0)_+^{-1}\left((F_1)_+^{-1}(D)\right)$;

(b) $(F_1 \circ F_0)_-^{-1}(D) = (F_0)_-^{-1}\left((F_1)_-^{-1}(D)\right)$.

Verify these relations!

Definition 1.2.11. Let X, Y_0, and Y_1 be sets, $F_0 : X \to P(Y_0)$, $F_1 : X \to P(Y_1)$ multimaps. The multimap $F_0 \times F_1 : X \to P(Y_0 \times Y_1)$,

$$(F_0 \times F_1)(x) = F_0(x) \times F_1(x)$$

is called *the Cartesian product of the multimaps F_0 and F_1.*

Lemma 1.2.12. *Let $D_0 \subset Y_0$, $D_1 \subset Y_1$, then*

(a) $(F_0 \times F_1)_+^{-1}(D_0 \times D_1) = (F_0)_+^{-1}(D_0) \cap (F_1)_+^{-1}(D_1)$;
(b) $(F_0 \times F_1)_-^{-1}(D_0 \times D_1) = (F_0)_-^{-1}(D_0) \cap (F_1)_-^{-1}(D_1)$.

Verify these relations!

1.2.2 *Upper and lover semicontinuity, continuity, closedness of multimaps*

Let X, Y be topological spaces, $F : X \to P(Y)$ multimaps.

Definition 1.2.13. A multimap F is said to be *upper semicontinuous at a point* $x \in X$ if for every open set $V \subset Y$ such that $F(x) \subset V$ there exists a neighborhood $U(x)$ of x such that

$$F(U(x)) \subset V.$$

Definition 1.2.14. A multimap F is called *upper semicontinuous (u.s.c.)* if it is upper semicontinuous at every point $x \in X$.

Consider some tantamount formulations.

Theorem 1.2.15. *The following conditions are equivalent:*

(a) *the multimap F is u.s.c.;*
(b) *for every open set $V \subset Y$, the set $F_+^{-1}(V)$ is open in X;*
(c) *for every closed set $W \subset Y$, the set $F_-^{-1}(W)$ is closed in X;*
(d) *if $D \subset Y$ then $F_-^{-1}(\overline{D}) \supset \overline{F_-^{-1}(D)}$.*

Proof. 1) The equivalence $(a) \Leftrightarrow (b)$ is evident;
2) the equivalence $(b) \Leftrightarrow (c)$ follows from Lemma 1.2.3(c) and Lemma 1.2.4(c);
3) $(c) \Rightarrow (d)$: $F_-^{-1}(\overline{D})$ is a closed set which contains $F_-^{-1}(D)$;
4) $(d) \Rightarrow (c)$: if D is closed then $F_-^{-1}(D) = F_-^{-1}(\overline{D}) \supset \overline{F_-^{-1}(D)}$, i.e., $F_-^{-1}(D)$ is closed. ■

Example 1.2.16. The multimaps from Examples 1.1.4 (a), (b); 1.1.5; 1.1.7 are u.s.c. The subdifferential multimap from Example 1.1.17 is also u.s.c. (see, e.g., [104]).

Definition 1.2.17. A multimap F is called *lower semicontinuous at a point* $x \in X$ if for every open set $V \subset Y$ such that $F(x) \cap V \neq \emptyset$ there exists a neighborhood $U(x)$ of x such that $F(x') \cap V \neq \emptyset$ for all $x' \in U(x)$.

Definition 1.2.18. A multimap F is said to be *lower semicontinuous (l.s.c.)* if it is lower semicontinuous at every point $x \in X$.

The lower semicontinuity also admits tantamount definitions.

Theorem 1.2.19. *The following conditions are equivalent:*

(a) *the multimap F is l.s.c.;*
(b) *for every open set $V \subset Y$, the set $F_-^{-1}(V)$ is open in X;*
(c) *for every closed set $W \subset Y$, the set $F_+^{-1}(W)$ is closed in X;*
(d) *if a system of open sets $\{V_j\}_{j \in J}$ forms a base for the topology of the space Y then for each V_j, the set $F_-^{-1}(V_j)$ is open in X;*
(e) *if $D \subset Y$ is an arbitrary set then $F_+^{-1}\left(\overline{D}\right) \supset \overline{F_+^{-1}(D)}$;*
(f) *if $A \subset X$ is an arbitrary set then $F\left(\overline{A}\right) \subset \overline{F(A)}$.*

Proof. 1) the equivalence $(a) \Leftrightarrow (b)$ is evident;
2) the equivalences $(b) \Leftrightarrow (c)$ and $(c) \Leftrightarrow (e)$ can be proved similarly to the corresponding statements of Theorem 1.2.15;
3) the equivalence $(b) \Leftrightarrow (d)$ follows from the fact that each set V_j is open and from Lemma 1.2.4 (d);
4) $(e) \Rightarrow (f)$: $F_+^{-1}\left(\overline{F(A)}\right) \supset \overline{F_+^{-1}(F(A))}$, but by virtue of Lemma 1.2.3(a) $F_+^{-1}(F(A)) \supset A$, hence $F_+^{-1}\left(\overline{F(A)}\right) \supset \overline{A}$. From Lemma 1.2.3$(b)$ it follows: $F\left(F_+^{-1}\left(\overline{F(A)}\right)\right) \subset \overline{F(A)}$, therefore $\overline{F(A)} \supset F\left(\overline{A}\right)$;
5) $(f) \Rightarrow (e)$: $F\left(\overline{F_+^{-1}(D)}\right) \subset \overline{F(F_+^{-1}(D))}$ but by virtue of Lemma 1.2.3 (b): $F\left(F_+^{-1}(D)\right) \subset D$, yielding $F\left(\overline{F_+^{-1}(D)}\right) \subset \overline{D}$. Applying F_+^{-1} to both sides of the last inclusion and using Lemma 1.2.3 (a) we get $F_+^{-1}\left(\overline{D}\right) \supset F_+^{-1}\left(F\left(\overline{F_+^{-1}(D)}\right)\right) \supset \overline{F_+^{-1}(D)}$.   ■

In the case of metric spaces we may obtain the following convenient sequential characterization of the lower and upper semicontinuity.

Theorem 1.2.20. *Let X and Y be metric spaces.*

(a) *For the lower semicontinuity of a multimap $F : X \to P(Y)$ at a point $x_0 \in X$ it is necessary and sufficient that:*

$(*)$ *for every sequence $\{x_n\}_{n=1}^{\infty} \subset X$, $x_n \to x_0$ and each $y_0 \in F(x_0)$ there exists a sequence $\{y_n\}_{n=1}^{\infty} \subset Y$, $y_n \in F(x_n)$ such that $y_n \to y_0$.*

(b) *For the upper semicontinuity of a multimap $F : X \to P(Y)$ at a point $x_0 \in X$ it is necessary, and in the case of the compactness of the set $F(x_0)$ it is also sufficient that:*

$(**)$ *for every sequences $\{x_n\}_{n=1}^{\infty} \subset X$, $x_n \to x_0$ and $\{y_n\}_{n=1}^{\infty} \subset Y$, $y_n \in F(x_n)$ the following relation holds: $\varrho_Y(y_n, F(x_0)) \to 0$.*

Proof. $(a)(i)$ Let condition $(*)$ holds. If the multimap F is not l.s.c. at the point x_0 then there exist an open set $V \subset Y$ such that $F(x_0) \cap V \neq \emptyset$ and a sequence $\{x_n\}_{n=1}^{\infty} \subset X$, $x_n \to x_0$ such that $F(x_n) \cap V = \emptyset$ for all $n = 1, 2, \ldots$ But these relations are in contradiction to the fact that we can, choosing a point $y_0 \in F(x_0) \cap V$, to find a sequence $y_n \in F(x_n)$ which converges to it.

$(a)(ii)$ Let a multimap F be l.s.c. at a point x_0 and a certain sequence $\{x_n\}_{n=1}^{\infty} \subset X$, $x_n \to x_0$ and a point $y_0 \in F(x_0)$ be given. Consider the sequence of open balls $B_{\frac{1}{m}}(y_0)$, $m = 1, 2, \ldots$ centered at the point y_0. Let a number n_1 be such that $F(x_n) \cap B_1(y_0) \neq \emptyset$ for all $n \geq n_1$. For every $n < n_1$ choose $y_n \in F(x_n)$ arbitrarily. Further, let us find a number $n_2 \geq n_1$ such that $F(x_n) \cap B_{\frac{1}{2}}(y_0) \neq \emptyset$ for all $n \geq n_2$. For every n, $n_1 \leq n < n_2$ choose $y_n \in F(x_n) \cap B_1(y_0)$. Continuing this process, we construct the desirable sequence y_n converging to y_0.

$(b)(i)$ Let a multimap F be u.s.c. at a point x_0. For an arbitrary $\varepsilon > 0$, consider an ε-neighborhood of the set $F(x_0)$:

$$U_\varepsilon(F(x_0)) = \{y \in Y \mid \varrho_Y(y, F(x_0)) < \varepsilon\}.$$

Since $U_\varepsilon(F(x_0))$ is the open set it follows that there exists a number n_ε such that for all $n \geq n_\varepsilon$ we get $F(x_n) \subset U_\varepsilon(F(x_0))$ implying $\varrho_Y(y_n, F(x_0)) \leq \varepsilon$.

$(b)(ii)$ Let $V \subset Y$ be an arbitrary open set containing $F(x_0)$. Then, by virtue of compactness of the set $F(x_0)$ there exists (see Chapter 0) $\varepsilon > 0$ such that $U_\varepsilon(F(x_0)) \subset V$. But then condition $(**)$ yields the existence of such a neighborhood $U(x_0)$ of the point x_0 such that $F(U(x_0)) \subset U_\varepsilon(F(x_0)) \subset V$. ∎

Definition 1.2.21. If a multimap F is upper and lower semicontinuous it is called *continuous*.

It is clear that in the case of a single-valued map, both upper, as well as lower semicontinuity mean usual continuity. Notice also that the constant multimap $F(x) \equiv Y_1 \subset Y$ is obviously continuous.

Example 1.2.22. (a) The multimaps from Examples 1.1.4 (a), (c); 1.1.5; 1.1.7; 1.1.8 are u.s.c. It can be verified by application of Definition 1.2.14 (**do it!**). Whence the multimaps from Examples 1.1.4 (a)(; 1.1.5; 1.1.7 are continuous. The multimap from Example 1.1.4 (b) is u.s.c., but not l.s.c., whereas the multimaps from Examples 1.1.4 (c)); 1.1.8 are l.s.c., but not u.s.c. In particular, for the multimap F from Example 1.1.8 we have $F_+^{-1}\left(\mathbb{R}_+^2\right) = \overline{\mathbb{R}}_+^2$, where

$$\mathbb{R}_+^2 = \left\{(x_1, x_2) \mid (x_1, x_2) \in \mathbb{R}^2, x_1 > 0, x_2 > 0\right\},$$

$$\overline{\mathbb{R}}_+^2 = \left\{(x_1, x_2) \mid (x_1, x_2) \in \mathbb{R}^2, x_1 \geq 0, x_2 \geq 0\right\}.$$

(b) Let T be a compact space; X a metric space, $C(T; X)$ denote the space of continuous functions endowed with the usual sup-norm. For an arbitrary nonempty subset $\Omega \subset C(T, X)$, the multimap $Q : T \to P(X)$ defined as

$$Q(t) = \Omega(t) := \{y(t) \mid y \in \Omega\}$$

is l.s.c. It can be checked up by using Theorem 1.2.20. **Verify** that if the set Ω is compact then the multimap Q is u.s.c. and hence continuous.

One more important class consists of closed multimaps.

Definition 1.2.23. A multimap F is called *closed* if its graph Γ_F (see Definition 1.1.2) is a closed subset of the space $X \times Y$.

Consider some tantamount formulations.

Theorem 1.2.24. *The following conditions are equivalent:*

(a) *the multimap F is closed;*
(b) *for each pair $x \in X, y \in Y$ such that $y \notin F(x)$ there exist neighborhoods $U(x)$ of x and $V(y)$ of y such that $F(U(x)) \cap V(y) = \emptyset$;*
(c) *for every nets $\{x_\alpha\} \subset X, \{y_\alpha\} \subset Y$ such that $x_\alpha \to x, y_\alpha \in F(x_\alpha), y_\alpha \to y$, we have $y \in F(x)$.*

Proof. 1) $(a) \Leftrightarrow (b)$: condition (b) means that a point $(x, y) \in X \times Y$ belongs to the complement of the graph Γ_F with a certain neighborhood; 2) $(a) \Leftrightarrow (c)$: condition (c) means that if a net $\{(x_\alpha, y_\alpha)\} \subset \Gamma_F$ converges to a point $(x, y) \in X \times Y$ then $(x, y) \in \Gamma_F$. ∎

Notice that in the case when X and Y are metric spaces, it is sufficient to consider in condition (c) usual sequences.

Example 1.2.25. The multimaps from Examples 1.1.4 (a), (b); 1.1.5 - 1.1.8 are closed.

Example 1.2.26. Consider Example 1.1.9. If X, Y are topological spaces, the space Y is Hausdorff and $f : X \to Y$ is a continuous surjective map, then the inverse multimap $F = f^{-1} : Y \to P(X)$ is closed.

Example 1.2.27. Consider Example 1.1.10. If X, Y, Z are topological spaces and the maps f and g are continuous then the implicit multimap F is closed.

Example 1.2.28. Consider Example1.1.11. If X, Y are topological spaces and function f is continuous then the multimap F_r is closed.

The validity of assertions in Examples 1.2.26 – 1.2.28 may be verified by applying Theorem 1.2.24(c) (**do it!**).

Introduce some notation which we will use in the sequel.

Let Y be a topological space.

Denote by $C(Y)$, $K(Y)$ the collections consisting of all nonempty closed, or respectively, compact subsets of Y. If the topological space Y is linear then $Pv(Y)$ denotes the collection of all nonempty convex subsets of Y. Introduce also the following symbols:

$$Cv(Y) = Pv(Y) \cap C(Y), \quad Kv(Y) = Pv(Y) \cap K(Y).$$

When a multimap F maps into the collections $C(Y)$, $K(Y)$ or $Pv(Y)$ we will say that F has closed, compact or convex values respectively.

From the definition of a closed multimap it follows that it has closed values.

The consideration of examples shows that closed and upper semicontinuous multimaps are a short distance apart. The relation between them is clarified by the following assertions.

Theorem 1.2.29. *Let X and Y be topological spaces. If the space Y is regular and a multimap $F : X \to C(Y)$ is u.s.c. then F is closed.*

Proof. Let $y \in Y, y \notin F(x)$. Since Y is regular there exist an open neighborhood $V(y)$ of the point y and an open set $W \supset F(x)$ such that $V(y) \cap W = \emptyset$. Let $U(x)$ be a neighborhood of x such that $F(U(x)) \subset W$. Then $F(U(x)) \cap V(y) = \emptyset$ and the statement follows from Theorem 1.2.24(b).   ∎

Remark 1.2.30. It is clear from the proof that when F has compact values the condition of regularity of Y can be replaced with the weaker condition that Y is a Hausdorff space.

To formulate a sufficient condition for a closed multimap to be u.s.c. we need the following definitions.

Definition 1.2.31. A multimap $F : X \to P(Y)$ is called:

   (a) *compact* if its range $F(X)$ is relatively compact in Y, i.e., $\overline{F(X)}$ is compact in Y;

   (b) *locally compact* if every point $x \in X$ has a neighborhood $U(x)$ such that the restriction F to $U(x)$ is compact;

   (c) *quasicompact* if the restriction of F to each compact subset $A \subset X$ is compact.

It is clear that $(a) \Rightarrow (b) \Rightarrow (c)$.

Theorem 1.2.32. *Let $F : X \to K(Y)$ be a closed multimap. If it is locally compact then it is u.s.c.*

Proof. Let $x \in X$, V an open set in Y such that $F(x) \subset V$. Let $U(x)$ be a neighborhood of x such that the restriction of F to it is compact and let $W = \overline{F(U(x))} \setminus V \neq \emptyset$. If $y \in W$ then since F is closed, there exist neighborhoods $V(y)$ of y and $U_y(x)$ of x such that $F(U_y(x)) \cap V(y) = \emptyset$. By the compactness of W we can extract a finite subcover $V(y_1), ..., V(y_n)$. Consider the following open neighborhood of x:

$$\widetilde{U}(x) = U(x) \cap \left(\bigcap_{i=1}^{n} U_{y_i}(x) \right).$$

Notice now that $x' \in \widetilde{U}(x)$ implies $F(x') \cap V(y_j) = \emptyset$ for all $j = 1, 2, ..., n$ and hence $F(x') \cap W = \emptyset$. From the other side, $F\left(\widetilde{U}(x)\right) \subset F(U(x))$. Therefore, $F\left(\widetilde{U}(x)\right) \subset V$.   ∎

The difference between closed and u.s.c. multimaps is illustrated by Examples 1.1.6–1.1.8. As it was mentioned already, the multimaps in these examples are closed, but they are not u.s.c. Notice that the multimap from Example 1.1.6 has compact values and the condition of its upper semicontinuity is violated at the same point $x = \pi/2$ in which the condition of the local compactness is not satisfied.

Let us consider some properties of closed and u.s.c. multimaps.

Theorem 1.2.33. *Let* $F : X \to C(Y)$ *be a closed multimap. If* $A \subset X$ *is a compact set then its image* $F(A)$ *is a closed subset of* Y.

Proof. The case $F(A) = Y$ is trivial. Let $y \in Y \setminus F(A)$. For any $x \in A$, let $U(x)$ and $V_x(y)$ be neighborhoods of x and y such that

$$F(U(x)) \cap V_x(y) = \emptyset.$$

If $U(x_1), ..., U(x_n)$ are neighborhoods forming a finite cover of A then $V(y) = \bigcap_{i=1}^{n} V_{x_i}(y)$ is a neighborhood of y such that $V(y) \cap F(A) = \emptyset$. ∎

Remark 1.2.34. The condition of compactness of the set A is essential: the image of a closed set under the action of a closed multimap can be a non-closed set. In fact, in Example 1.1.7: $F(\overline{\mathbb{R}_+}) = (0, 1]$.

In the sequel an important role will be played by the following property of u.s.c. multimaps.

Theorem 1.2.35. *Let* $F : X \to K(Y)$ *be a u.s.c. multimap. If* $A \subset X$ *is a compact set then its image* $F(A)$ *is a compact subset of* A.

Proof. Let $\{V_j\}_{j \in J}$ be an open cover of the set $F(A)$. For each point $x \in A$, the value $F(x)$ can be covered by a finite collection of sets $V_{j_1}, ..., V_{j_{n(x)}}$. We denote $V_x = \cup_{i=1}^{n(x)} V_{j_i}$. The sets $F_+^{-1}(V_x)$, $x \in A$ form an open cover of A. If we select a finite subcover

$$F_+^{-1}(V_{x_1}), .., F_+^{-1}(V_{x_m})$$

then the sets $V_{x_1}, ..., V_{x_m}$ form an open cover of the set $F(A)$. ∎

Remark 1.2.36. The condition of a upper semicontinuity is essential in this theorem. In fact, for a closed multimap F with compact values in Example 1.1.6 we have $F([0, \pi]) = \mathbb{R}$.

Let us mention also the following property.

Theorem 1.2.37. *Let X and Y be topological spaces; $A \subset X$ a connected set and $F : X \to P(Y)$ a multimap. If one of the following conditions holds true:*

> *(i) F is upper or lower semicontinuous and the values $F(x)$ are connected for each $x \in A$;*
> *(ii) F is continuous and a value $F(x_0)$ is connected for some $x_0 \in A$*

then $F(A)$ is a connected subset of Y.

Proof. (i) Consider the case of a upper semicontinuous multimap F. Suppose the contrary, then there exist open sets V_0 and V_1 in the space Y such that:

> $a)$ $F(A) \subset (V_0 \cup V_1)$;
> $b)$ $F(A) \cap V_i \neq \emptyset, \ i = 0, 1$;
> $c)$ $(F(A) \cap V_0) \cap (F(A) \cap V_1) = \emptyset$.

Consider the sets $F_+^{-1}(V_0) = U_0$, $F_+^{-1}(V_1) = U_1$. These sets are open by virtue of the upper semicontinuity of F. Notice that the value of each point $x \in A$ is contained only in one of the sets V_0 or V_1 since otherwise the set $F(x)$ would be disconnected. Hence $A \subset (U_0 \cup U_1)$; $A \cap U_i \neq \emptyset$ for each $i = 0, 1$ and $(A \cap U_0) \cap (A \cap U_1) = \emptyset$ that contradicts to the connectedness of the set A.

In the case when the multimap F is lower semicontinuous, it is sufficient to note that open sets arising in the definition of a connected set may be replaced with closed ones and to carry out the same reasonings as above, by using Theorem 1.2.19 (c).

(ii) Also suppose the contrary. Then, by virtue of its connectedness, the set $F(x_0)$ must lie either in V_0 or V_1. Suppose for determinacy that $F(x_0) \subset V_0$ and hence $F_+^{-1}(V_0) \neq \emptyset$. Then we get

$$A \subset F_+^{-1}(V_0) \bigcup F_-^{-1}(V_1),$$

and moreover, by the continuity of the multimap F, both last sets are non-empty, disjoint and open. But this contradicts to the connectedness of A. ∎

1.2.3 *Multivalued maps into a metric space*

In the case when a multimap acts into a metric space we can obtain a few convenient characterizations for the above considered types of continuity.

Everywhere in this section, (Y, ϱ) is a metric space.

Definition 1.2.38. Let $F : X \to P(Y)$ be a multimap. The multimap $F_\varepsilon : X \to P(Y)$,

$$F_\varepsilon(x) = U_\varepsilon(F(x)) = \{y | y \in Y, \ \varrho(y, F(x)) < \varepsilon\}$$

is called an *ε-enlargement of the multimap F.*

Theorem 1.2.39. *For the upper semicontinuity of a multimap $F : X \to K(Y)$ at a point $x \in X$, it is necessary and sufficient that for every $\varepsilon > 0$ there exists a neighborhood $U(x)$ of x such that $F(x') \subset F_\varepsilon(x)$ for all $x' \in U(x)$.*

Proof. *1) Necessity.* Notice that

$$F_\varepsilon(x) = \bigcup_{y \in F(x)} B_\varepsilon(y)$$

is an open set containing $F(x)$ and apply Definition 1.2.13.

2) Sufficiency. Let $F(x) \subset V$, where V is an open set. Then (see Ch. 0) there exists $\varepsilon > 0$ such that $F_\varepsilon(x) \subset V$. But then there exists a neighborhood $U(x)$ of x such that $F(U(x)) \subset F_\varepsilon(x) \subset V$. ∎

Theorem 1.2.40. *For the lower semicontinuity of a multimap $F : X \to K(Y)$ at a point $x \in X$, it is necessary and sufficient that for every $\varepsilon > 0$ there exists a neighborhood $U(x)$ of x that $F(x) \subset F_\varepsilon(x')$ for all $x' \in U(x)$.*

Proof. *1) Necessity.* Take $\varepsilon > 0$ and let $y_1, ..., y_n$ be points of the set $F(x)$ such that the collection of balls $B_{\frac{\varepsilon}{2}}(y_i), 1 \le i \le n$ forms an open cover of $F(x)$. Since F is l.s.c., for every $i, 1 \le i \le n$, there exists an open neighborhood $U_i(x)$ of the point x such that from $x' \in U_i(x)$ it follows that $F(x') \cap B_{\frac{\varepsilon}{2}}(y_i) \ne \emptyset$. But then, $x' \in U(x) = \bigcap_{i=1}^{n} U_i(x)$ implies $B_{\frac{\varepsilon}{2}}(y_i) \subset F_\varepsilon(x')$ for all $i, 1 \le i \le n$ and hence the neighborhood $U(x)$ is the desired one.

2) Sufficiency. Let V be an open set in Y and $F(x) \cap V \ne \emptyset$. Take an arbitrary point $y \in F(x) \cap V$ and let $\varepsilon > 0$ be such that $B_\varepsilon(y) \subset V$. Let $U(x)$ be a neighborhood of x such that $x' \in U(x)$ implies $F(x) \subset F_\varepsilon(x')$. Then $F(x') \cap B_\varepsilon(y) \ne \emptyset$ for all $x' \in U(x)$ proving that F is l.s.c. at x. ∎

It is worth noting that in the necessary part of Theorem 1.2.39 and in the sufficient part of Theorem 1.2.40 the compactness of the values of the multimap F is not used.

As earlier, let $C(Y)$ denote the collection of all nonempty closed subsets of Y. For $A, B \in C(Y)$, the value

$$\varrho_* (A, B) = \sup_{a \in A} \varrho(a, B)$$

is called *the deviation of the set A from the set B*. The function $\varrho_* : C(Y) \times C(Y) \to \mathbb{R} \cup \{\infty\}$ possesses the following properties.

Theorem 1.2.41.

    (a) $\varrho_* (A, B) \geq 0$ *for each* $A, B \in C(Y)$;

    (b) $\varrho_* (A, B) = 0$ *implies* $A \subset B$;

    (c)   *in a general case* $\varrho_* (A, B) \neq \varrho_* (B, A)$;

    (d)   *if* $\varrho_* (A, B) < \infty$ *then* $\varrho_* (A, B) \leq \varrho_* (A, C) + \varrho_* (C, B)$ *for every* $C \in C(Y)$;

    (e)   *if* $\varrho_* (A, B) < \infty$ *then* $\varrho_* (A, B) = \inf \{\varepsilon \mid A \subset U_\varepsilon (B)\}$.

Proof. (a) Follows immediately from the definition.

(b) For each $x \in A$ we have $\varrho(x, B) = 0$. Hence x is a limit point for a certain sequence of points from B. Since B is closed, we get $x \in B$.

(c) Take $A = \{a\} \in Y$, $B = \{a\} \cup \{b\} \in Y$, $a \neq b$. Then $\varrho_* (A, B) = 0$, $\varrho_* (B, A) = \varrho(b, a) \neq 0$.

(d) By the triangle inequality, for each $x \in A$ we have

$$\varrho(x, B) = \inf_{y \in B} \varrho(x, y) \leq \inf_{y \in B} (\varrho(x, z) + \varrho(z, y)) = \varrho(x, z) + \varrho(z, B),$$

where z is an arbitrary point of C. Then

$$\varrho(x, B) \leq \varrho(x, z) + \varrho_* (C, B)$$

for each $z \in C$. Whence

$$\varrho(x, B) \leq \inf_{z \in C} \varrho(x, z) + \varrho_* (C, B) = \varrho(x, C) + \varrho_* (C, B)$$

$$\leq \varrho_* (A, C) + \varrho_* (C, B).$$

Then $\varrho_* (A, B) \leq \varrho_* (A, C) + \varrho_* (C, B)$.

(e) Let $\varepsilon > \varrho_* (A, B)$, then for each point $x \in A$ there exists a point $y \in B$ such that $x \in B_\varepsilon (y)$. Therefore $A \subset U_\varepsilon (B)$, i.e., $\inf \{\varepsilon \mid A \subset U_\varepsilon (B)\} \leq \varrho_* (A, B)$. In case when $\varepsilon > 0$ is such that $A \subset U_\varepsilon (B)$, for every $x \in A$ we have $\varrho(x, B) < \varepsilon$. Then $\varrho_* (A, B) \leq \varepsilon$, i.e., $\varrho_* (A, B) \leq \inf \{\varepsilon \mid A \subset U_\varepsilon (B)\}$. Comparing the obtained inequalities we get the desired property. ∎

Consider the function $h : C(Y) \times C(Y) \to \mathbb{R} \cup \{\infty\}$,

$$h(A, B) = \max\{\varrho_*(A, B), \ \varrho_*(B, A)\}.$$

Applying the previous result one can verify **(do it!)** that this function has the next properties:

For each $A, B \in C(Y)$ the following holds true:

1) $h(A, B) \geq 0$;
2) $h(A, B) = 0$ *is equivalent to* $A = B$;
3) $h(A, B) = h(B, A)$;
4) *If* $h(A, B) < \infty$ *then* $h(A, B) \leq h(A, C) + h(C, B)$ *for each* $C \in C(Y)$.

Definition 1.2.42. The function h is called *the extended Hausdorff metric* on the set $C(Y)$.

Here the term "extended" means that the function h can take infinite values.

Denote $Cb(Y)$ *the collection of all nonempty closed bounded subsets of* Y. From the above properties it immediately follows that the function h is a usual metric on this set. It is called *the Hausdorff metric*.

Notice that from Theorem 1.2.41(e) it follows that for every $A, B \in Cb(Y)$ the Hausdorff metric may be defined as

$$h(A, B) = \inf\{\varepsilon | A \subset U_\varepsilon(B), B \subset U_\varepsilon(A)\}.$$

Definition 1.2.43. A multimap $F : X \to Cb(Y)$ is called *Hausdorff continuous*, if it is continuous as a single-valued map into the metric space $(Cb(Y), h)$.

For multimaps with compact values we can obtain now the following useful characterization of the continuity.

Theorem 1.2.44. *A multimap $F : X \to K(Y)$ is continuous if and only if it is Hausdorff continuous.*

Proof. The statement of the theorem directly follows from Theorems 1.2.39 and 1.2.40. ∎

Now, let Y be a separable metric space. The following criteria of lower and upper semicontinuity of multimaps will be useful in the sequel.

Let $\{r_n\}_{n=1}^{\infty}$ be a countable dense subset of Y. For a multimap $F : X \to P(Y)$ define the functions $\{\varphi_n\}_{n=1}^{\infty}$, $\varphi_n : X \to \mathbb{R}$,

$$\varphi_n(x) = \varrho(r_n, F(x)).$$

Theorem 1.2.45. *For the lower semicontinuity of a multimap $F : X \to P(Y)$ it is necessary and sufficient that all the functions φ_n are upper semicontinuous (in the single-valued sense).*

Proof. For each $a > 0$ and n the set

$$\{x \mid x \in X, \varphi_n(x) < a\}$$

coincides with the set $F_{-}^{-1}(B_a(r_n))$. To verify the assertion it remains to notice that the balls centered at the points r_n form the base of the topology of the space Y and to use Theorem 1.2.19 (d). ∎

For the further reasonings we need the following statement.

Lemma 1.2.46. *Let $F : X \to P(Y)$ be a multimap, $W \subset Y$ a closed subset. Let sets W_m be such that $W \subset W_m \subset U_{\varepsilon_m}(W)$ for a certain sequence $\{\varepsilon_m\}_{m=1}^{\infty}, \varepsilon_m > 0, \varepsilon_m \to 0$. Then:*

(a) $F_{+}^{-1}(W) = \bigcap_{m=1}^{\infty} F_{+}^{-1}(W_m);$

(b) if the values of the multimap F are compact then

$$F_{-}^{-1}(W) = \bigcap_{m=1}^{\infty} F_{-}^{-1}(W_m).$$

Proof. The inclusions

$$F_{+}^{-1}(W) \subset \bigcap_{m=1}^{\infty} F_{+}^{-1}(W_m), \quad F_{-}^{-1}(W) \subset \bigcap_{m=1}^{\infty} F_{-}^{-1}(W_m)$$

are evident.

(a) If $x \notin F_{+}^{-1}(W)$ then there exists $y \in F(x)$ such that $y \notin W$. But then $\varepsilon_m < \varrho(y, W)$ yields $x \notin F_{+}^{-1}(W_m)$ that proves (a).

(b) If $x \notin F_{-}^{-1}(W)$ then $F(x) \cap W = \emptyset$ and, since the set $F(x)$ is compact there exists $\varepsilon > 0$ such that $F_{\varepsilon}(x) \cap W = \emptyset$. But then $\varepsilon_m < \varepsilon$ implies $x \notin F_{-}^{-1}(W_m)$ and (b) is also proved. ∎

Theorem 1.2.47. *For the upper semicontinuity of a multimap* $F : X \to K(Y)$ *it is necessary and, in the case of compactness of the multimap F, also sufficient that all the functions φ_n are lower semicontinuous (in the single-valued sense).*

Proof. 1) *Necessity.* If the multimap F is upper semicontinuous then for each $a > 0$ and n the set

$$\{x \mid x \in X, \varphi_n(x) > a\} = F_+^{-1}\left(Y \setminus \overline{B}_a(r_n)\right)$$

is open.

2) *Necessity.* Since the multimap F is compact, it is sufficient to show that for every compact set $K \subset Y$ the set $F_-^{-1}(K)$ is closed. For a certain sequence $\{\varepsilon_m\}_{m=1}^{\infty}$, $\varepsilon_m > 0$, $\varepsilon_m \to 0$ consider the finite covers of K by closed balls of the radius ε_m centered at points from the set $\{r_n\}_{n=1}^{\infty}$:

$$K \subset K_m = \bigcup_{i=1}^{q(m)} \overline{B}_{\varepsilon_m}\left(r_{n_{(m,i)}}\right).$$

For each m, the set

$$F_-^{-1}(K_m) = \bigcup_{i=1}^{q(m)} F_-^{-1}\left(\overline{B}_{\varepsilon_m}\left(r_{n_{(m,i)}}\right)\right) = \bigcup_{i=1}^{q(m)} \left\{x \mid x \in X, \varphi_{n_{(m,i)}}(x) \leq \varepsilon_m\right\}$$

is closed. Applying Lemma 1.2.46 *(b)* we obtain $F_-^{-1}(K) = \bigcap_{m=1}^{\infty} F_-^{-1}(K_m)$ from where the closedness of the set $F_-^{-1}(K)$ follows. ∎

In conclusion of this section notice that for metric spaces we have the following refinement of Theorem 1.2.32.

Theorem 1.2.48. *Let X and Y be metric spaces and $F : X \to K(Y)$ a closed quasicompact multimap Then F is upper semicontinuous.*

Proof. Let $x \in X$ be a point and $V \subset Y$ an open set such that $F(x) \subset V$. If F is not u.s.c. at x there exists a sequence $\{x_n\} \subset X$, $x_n \to x$ such that we can choose a sequence $y_n \in F(x_n) \setminus V$ for all $n = 1, 2, \ldots$ By virtue of the quasicompactness condition we can assume without loss of generality that $y_n \to y \notin V$, contrary to $y \in F(x)$. ∎

1.3 Operations on multivalued maps

In mathematics there are no symbols for obscure thoughts.

—*Henri Poincaré*

The variety of operations that can be defined on multimaps is intrinsically richer than for single-valued maps: such operations as union, intersection of multimaps and some others have no "single-valued" analogs. In this section we investigate the preserving of continuity properties of multimaps with respect to various operations on them.

1.3.1 *Set-theoretic operations*

Let X, Y be topological spaces; $\{F_j\}_{j \in J}$, $F_j : X \to P(Y)$ a family of multimaps.

Theorem 1.3.1. (a) *Let multimaps F_j be upper semicontinuous. If the set of indices J is finite then the union of multimaps $\bigcup\limits_{j \in J} F_j : X \to P(Y)$,*

$$\left(\bigcup_{j \in J} F_j \right)(x) = \bigcup_{j \in J} F_j(x)$$

is upper semicontinuous;

(b) *Let the multimaps F_j be lower semicontinuous. Then their union $\bigcup\limits_{j \in J} F_j$ is lower semicontinuous;*

(c) *Let multimaps $F_j : X \to C(Y)$ be closed. If the set of indices J is finite then the union $\bigcup\limits_{j \in J} F_j : X \to C(Y)$ is closed.*

Proof. (a) Let $V \subset Y$ be open, then in accordance with Lemma 1.2.7 (a)

$$\left(\bigcup_{j \in J} F_j \right)_+^{-1}(V) = \bigcap_{j \in J} (F_j)_+^{-1}(V),$$

and hence this set is open and by Theorem 1.2.15 (b) the multimap $\bigcup\limits_{j \in J} F_j$ is u.s.c.

(b) The assertion similarly follows from Lemma 1.2.8 (a) and Theorem 1.2.19 (b).

(c) It is easy to verify (**do it!**) that the graph $\Gamma_{\underset{j \in J}{\bigcup} F_j}$ of the multimap $\underset{j \in J}{\bigcup} F_j$ is the union of the graphs $\underset{j \in J}{\bigcup} \Gamma_{F_j}$ from where the asserion follows. ∎

Theorem 1.3.2. (a) *Let multimaps* $F_j : X \to C(Y)$ *be upper semicontinuous. If the set of indices J is finite, the space Y is normal and* $\underset{j \in J}{\bigcap} F_j(x) \neq \emptyset, \forall x \in X$ *then the intersection of multimaps* $\underset{j \in J}{\bigcap} F_j : X \to C(Y)$,

$$\left(\bigcap_{j \in J} F_j \right)(x) = \bigcap_{j \in J} F_j(x)$$

is upper semicontinuous.

(b) *Let multimaps* $F_j : X \to C(Y)$ *be closed and* $\underset{j \in J}{\bigcap} F_j(x) \neq \emptyset, \forall x \in X$. *then the intersection* $\underset{j \in J}{\bigcap} F_j : X \to C(Y)$ *is closed.*

Proof. (a) At first, let us prove the assertion for the case of two multimaps F_0 and F_1. For $x \in X$, let V be a neighborhood of the set $F(x) = (F_0 \cap F_1)(x)$. If at least one of the sets $F_0(x)$ or $F_1(x)$ is contained in V then the existence of such a neighborhood $U(x)$ of x that $F(U(x)) \subset V$ is evident. Otherwise $F_0(x) \backslash V$ and $F_1(x) \backslash V$ are nonempty disjoint closed sets. By virtue of the normality of the space Y there exist disjoint open sets W_0 and W_1 such that $(F_j(x) \backslash V) \subset W_j, j = 0, 1$. Then for every $j = 0, 1$ we have

$$F_j(x) \subset \left(V \cup (F_j(x) \backslash V)\right) \subset (V \cup W_j).$$

From the upper semicontinuity of the multimaps F_j it follows that for each $j = 0, 1$ there exists a neighborhood $U_j(x)$ of x such that

$$F_j(U_j(x)) \subset (V \cup W_j).$$

But if $U(x) = U_0(x) \cap U_1(x)$ then for every $x' \in U(x)$ we have

$$F(x') = F_0(x') \cap F_1(x') \subset \left((V \cup W_0) \cap (V \cup W_1)\right) = V$$

proving the upper semicontinuity of F at x.

The validity of the statement in the general case now follows from the mathematical induction principle.

(b) The statement follows from the fact that the graph of the multimap $\underset{j \in J}{\bigcap} F_j$ is the intersection of the graphs of the multimaps F_j. ∎

Let us mention also the following assertion.

Theorem 1.3.3. *Let a multimap $F_0 : X \to C(Y)$ be closed, a multimap $F_1 : X \to K(Y)$ upper semicontinuous and*

$$F_0(x) \cap F_1(x) \neq \emptyset, \forall x \in X.$$

Then the intersection $F = F_0 \cap F_1 : X \to K(Y)$ is upper semicontinuous.

Proof. For an arbitrary $x \in X$ let $V \subset Y$ be any open neighborhood of the set $(F_0 \cap F_1)(x)$. We will show that there exists an open neighborhood $U(x)$ of x such that $(F_0 \cap F_1)(U(x)) \subset V$.

When $F_1(x) \subset V$ the existence of such a neighborhood follows from the upper semicontinuity of F_1. If $K = F_1(x) \setminus V \neq \emptyset$ then the set K is compact and $K \cap F_0(x) = \emptyset$. As F_0 is a closed multimap, for each point $y \in K$ there exist neighborhoods $V(y) \subset Y$ of y and $U_y(x) \subset X$ of x such that $F_0(U_y(x)) \cap V(y) = \emptyset$ (see Theorem 1.2.24(b)).

Let $\{V(y_i)\}_{i=1}^{n}$ be a finite cover of K formed by such neighborhoods $V(y)$, and $V(K) = \bigcup_{i=1}^{n} V(y_i)$. The open set $V \cup V(K)$ contains $F_1(x)$, hence there exists a neighborhood $U_1(x)$ of x such that $F_1(U_1(x)) \subset (V \cup V(K))$. Then the neighborhood

$$U(x) = U_1(x) \cap U_{y_1}(x) \cap \ldots \cap U_{y_n}(x)$$

is the required one. In fact, $F_0(U(x)) \cap V(K) = \emptyset$ and $F_1(U(x)) \subset (V \cup V(K))$, therefore $(F_0 \cap F_1)(U(x)) \subset V$. ∎

Corollary 1.3.4. *Let a multimap $F : X \to K(Y)$ be upper semicontinuous, $C \subset Y$ a closed set and $F(x) \cap C \neq \emptyset, \forall x \in X$. Then the multimap $\tilde{F} : X \to K(Y)$,*

$$\tilde{F}(x) = F(x) \cap C$$

is upper semicontinuous.

Proof. It is clear that the multimap $F_0 : X \to C(Y)$,

$$F_0(x) \equiv C$$

is closed. Take $F_1 = F$ and apply the previous theorem. ∎

Corollary 1.3.5. *Let Y be a Hausdorff topological space, multimaps $\{F_j\}_{j \in J}, F_j : X \to K(Y)$ upper semicontinuous and $\bigcap_{j \in J} F_j(x) \neq \emptyset, \forall x \in X$. Then the intersection $\bigcap_{j \in J} F_j : X \to K(Y)$ is upper semicontinuous.*

Proof. Let F_{j_0} be one of the multimaps from the family. Since al the multimaps F_j are closed (see Remark 1.2.30) the multimap

$$F_1 = \bigcap_{j \in J, j \neq j_0} F_j$$

is also closed (Theorem 1.3.2(b)). Take $F_0 = F_{j_0}$ and apply the theorem. ∎

The continuity properties of the intersection of lower semicontinuous multimaps are more complicated. The following example demonstrates that in general case such an intersection is not lower semicontinuous.

Example 1.3.6. Consider the multimaps $F_0, F_1 : [0, \pi] \to Kv\left(\mathbb{R}^2\right)$ defined in the following way. The multimap F_0 is constant:

$$F_0\left(\varphi\right) = \left\{(y_0, y_1) \mid (y_0, y_1) \in \mathbb{R}^2, y_1 \geq 0, y_0^2 + y_1^2 \leq 1\right\},$$

whereas the multimap F_1 is defined as

$$F_1\left(\varphi\right) = \left\{(y_0, y_1) \mid (y_0, y_1) \in \mathbb{R}^2, y_0 = \lambda \cos \varphi, y_1 = \lambda \sin \varphi, \lambda \in [-1, 1]\right\}$$

(See fig. 8).

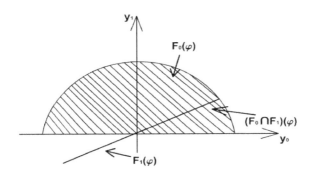

Fig. 8

The multimaps F_0 and F_1 are l.s.c. (they are even continuous) but their intersection $F_0 \cap F_1$ defined on the whole interval $[0, \pi]$ loses the lower semicontinuity property at the points 0 and π **(why?)**

To clarify the conditions under which we can guarantee the lower semicontinuity of the intersection of multimaps the following notion is useful.

Definition 1.3.7. A multimap $F : X \to P\left(Y\right)$ is called *quasi-open at a point* $x \in X$ if

$$intF\left(x\right) \neq \emptyset$$

and for every $y \in intF(x)$ there exist neighborhoods $V(y) \subset Y$ of y and $U(x) \subset X$ of x such that $V(y) \subset F(x')$ for all $x' \in U(x)$. A multimap F is said to be *quasi-open* provided it is quasi-open at every point $x \in X$.

It is easy to see that a multimap $F : X \to P(Y)$ such that $intF(x) \neq \emptyset$ for all $x \in X$ is quasi-open if and only if the multimap $intF : X \to P(Y)$,

$$(intF)(x) = intF(x)$$

has the open graph $\Gamma_{intF} \subset X \times Y$.

We have the following important characterization of a quasi-open multimap.

Theorem 1.3.8. *Let Y be a finite-dimensional linear topological space. A multimap $F : X \to Cv(Y)$ is quasi-open at a point $x \in X$ if and only if $intF(x) \neq \emptyset$ and F is lower semicontinuous at x.*

Proof. 1) Let F be quasi-open at $x \in X$ then $intF(x) \neq \emptyset$. If $V \subset Y$ is an open set such that $V \cap F(x) \neq \emptyset$ then it is easy to see that $V \cap intF(x) \neq \emptyset$. For an arbitrary $y \in Y \cap intF(x)$ let $V(y) \subset Y$ and $U(x) \subset X$ be neighborhoods such that $V(y) \subset F(x')$ for all $x' \in U(x)$. But $V(y) \cap V \neq \emptyset$ implies that $V \cap F(x') \neq \emptyset$ for all $x' \in U(x)$, giving the lower semicontinuity of F at x.

2) Conversely, let $intF(x) \neq \emptyset$ and F be l.s.c. at $x \in X$. Let $y \in intF(x)$ and $B_\delta(y) \subset F(x)$ for some $\delta > 0$. Take $\delta_1, 0 < \delta_1 < \delta$. Since the space Y is finite-dimensional, the ball $B_\delta(y)$ is relatively compact. By applying the reasonings similar to those that were used while proving the necessity part of Theorem 1.2.40 we get that there exists a neighborhood $U(x)$ of x such that for each point $x' \in U(x)$ we have $B_\delta(y) \subset F_\eta(x')$, where $\eta = \delta - \delta_1$.

Let now $y' \in B_{\delta_1}(y)$ but $y' \notin F(x')$ for some $x' \in U(x)$. Then from the convexity of the set $F(x')$ it follows that the ball $B_\eta(y')$ will contain points whose distance from $F(x')$ is greater than η. But this contradicts to the fact that $B_\eta(y') \subset B_\delta(y) \subset F_\eta(x')$. Therefore, $B_{\delta_1}(y) \subset F(x')$ for all $x' \in U(x)$. $\blacksquare$

We now formulate a condition that guarantees the lower semicontinuity of the intersection of multimaps.

Theorem 1.3.9. *Let X, Y be topological spaces; a multimap $F_0 : X \to P(Y)$ be lower semicontinuous at $x_0 \in X$ and a multimap $F_1 : X \to P(Y)$*

be quasi-open at x_0, and

$$F_0(x) \cap F_1(x) \neq \emptyset$$

for all $x \in X$. If

$$F_0(x_0) \cap F_1(x_0) \subset \overline{F_0(x_0) \cap \mathrm{int} F_1(x_0)},$$

then the intersection $F_0 \bigcap F_1$ is lower semicontinuous at x_0.

Proof. Let $V \subset Y$ be an open set such that $V \cap (F_0 \cap F_1)(x_0) \neq \emptyset$. From the assumptions it follows that there exists a point $y \in V \cap (F_0 \cap F_1)(x_0)$ which is an interior point of the set $F_1(x_0)$. Let $V(y)$ be a neighborhood of y such that $V(y) \subset (V \cap F_1(x_0))$. By using the quasi-openness of the multimap F_1 we can assume, without loss of generality, that there exists a neighborhood $U_1(x_0)$ of x_0 such that $V(y) \subset F_1(x')$ for all $x' \in U_1(x_0)$.

Since $y \in F_0(x_0)$ and the multimap F_0 is l.s.c. there exists a neighborhood $U_0(x_0)$ of x_0 such that $F_0(x'') \cap V(y) \neq \emptyset$ for all $x'' \in U_0(x_0)$. But then for every $\tilde{x} \in U(x_0) = U_0(x_0) \cap U_1(x_0)$ we have $(F_0 \cap F_1)(\tilde{x}) \cap V(y) \neq \emptyset$, i.e., $(F_0 \cap F_1)(\tilde{x}) \cap V \neq \emptyset$, that means the lower semicontinuity of the multimap $F_0 \cap F_1$ at x_0. ∎

As a corollary, we can obtain now a sufficient condition for the lower semicontinuity of the intersection of l.s.c. multimaps.

Theorem 1.3.10. *Let X be a topological space, Y a finite-dimensional linear topological space, $F_0, F_1 : X \to Cv(Y)$ l.s.c. multimaps. Assume that $F_0(x) \cap F_1(x) \neq \emptyset$ for all $x \in X$ and*

$$F_0(x_0) \cap \mathrm{int} F_1(x_0) \neq \emptyset$$

for some $x_0 \in X$. Then the intersection $F_0 \cap F_1 : X \to Cv(Y)$ is l.s.c. at x_0.

Proof. From Theorem 1.3.8 it follows that the multimap F_1 is quasi-open at x_0. Let $y \in (F_0 \cap F_1)(x_0)$ be an arbitrary point and $\tilde{y} \in F_0(x_0) \cap \mathrm{int} F_1(x_0)$.

It is clear that $[y, \tilde{y}] \subset (F_0 \cap F_1)(x_0)$ and $(y, \tilde{y}] \subset \mathrm{int} F_1(x_0)$. This means that the multimaps F_0 and F_1 satisfy the assumptions of Theorem 1.3.9. ∎

It is worth noting that the loss of the lower semicontinuity for the intersection of multimaps in Example 1.3.6 occurs exactly at the points where the above condition is violated.

Now consider some continuity properties of the composition of multimaps (see Definition 1.2.9).

Let X, Y, and Z be topological spaces.

Theorem 1.3.11. *If the multimaps $F_0 : X \to P(Y)$ and $F_1 : Y \to P(Z)$ are u.s.c. (l.s.c.) then their composition $F_1 \circ F_0 : X \to P(Z)$ is u.s.c. (respectively, l.s.c.).*

Proof. The assertion follows immediately from Theorems 1.2.15(b), 1.2.19(b) and Lemma 1.2.10. ∎

Theorem 1.3.12. *Let $F_0 : X \to K(Y)$ be a u.s.c. multimap and $F_1 : Y \to C(Z)$ a closed multimap. Then the composition $F_1 \circ F_0 : X \to C(Z)$ is a closed multimap.*

Proof. Let $z \in Z$ be such that $z \notin F_1 \circ F_0(x)$, $x \in X$. Applying Theorem 1.2.24(b) to the closed multimap F_1 we can find for each point $y \in F_0(x)$, neighborhoods $W_y(z)$ of z and $V(y)$ of y such that

$$F_1(V(y)) \cap W_y(z) = \emptyset.$$

Let $\{V(y_i)\}_{i=1}^n$ be a finite cover of the set $F_0(x)$. If now $U(x)$ is a neighborhood of x such that

$$F_0(U(x)) \subset \bigcup_{i=1}^n V(y_i)$$

then

$$(F_1 \circ F_0)(U(x)) \cap \left(\bigcap_{i=1}^n W_{y_i}(z) \right) = \emptyset,$$

and the application of Theorem 1.2.24(b) concludes the proof. ∎

Remark 1.3.13. The condition of upper semicontinuity of the multimap F_0 is essential. The following example shows that the composition of closed multimaps is not necessarily a closed multimap.

Example 1.3.14. The multimaps $F_0 : \mathbb{R} \to K(\mathbb{R})$,

$$F_0(x) = \begin{cases} \{\frac{1}{x}\}, & x \neq 0 \\ \{0\}, & x = 0 \end{cases}$$

and $F_1 : \mathbb{R} \to K(\mathbb{R})$,

$$F_1(x) = \begin{cases} \{\frac{1}{x}\}, & x \neq 0 \\ \{1\}, & x = 0 \end{cases}$$

are closed but not u.s.c. Their composition $F_1 \circ F_0 : \mathbb{R} \to K(\mathbb{R})$,

$$(F_1 \circ F_0)(x) = \begin{cases} \{x\}, x \neq 0 \\ \{1\}, x = 0 \end{cases}$$

is not closed.

We consider now the Cartesian product of multimaps (see Definition 1.2.11).

Theorem 1.3.15. *If multimaps $F_0 : X \to P(Y)$, $F_1 : X \to P(Z)$ are lower semicontinuous then their Cartesian product $F_0 \times F_1 : X \to P(Y \times Z)$ is lower semicontinuous.*

Proof. Notice that the sets $V_0 \times V_1$, where $V_0 \subset Y$, $V_1 \subset Z$ are open sets form a base for the topology of the space $Y \times Z$ and apply Theorem 1.2.19(d) and Lemma 1.2.12(b). ∎

Theorem 1.3.16. *If multimaps $F_0 : X \to C(Y)$, $F_1 : X \to C(Z)$ are closed then their Cartesian product $F_0 \times F_1 : X \to C(Y \times Z)$ is closed.*

Proof. Consider nets $\{x_\alpha\} \subset X$, $\{v_\alpha\} \subset Y \times Z$ such that $x_\alpha \to x$, $v_\alpha \in (F_0 \times F_1)(x_\alpha)$, $v_\alpha \to v$. Then $v_\alpha = y_\alpha \times z_\alpha$, $y_\alpha \in F_0(x_\alpha)$, $z_\alpha \in F_1(x_\alpha)$. from the definition of the topology in $Y \times Z$, the convergence $v_\alpha \to v = (y, z)$ implies the convergences $y_\alpha \to y$ and $z_\alpha \to z$. From the closedness of the multimaps F_0 and F_1 it follows that $y \in F_0(x)$, $z \in F_1(x)$ (Theorem 1.2.24(c)) but it means that $v \in (F_0 \times F_1)(x)$ concluding the proof. ∎

To consider the upper semicontinuity of the Cartesian product of multimaps we need the compactness of their values.

Theorem 1.3.17. *If multimaps $F_0 : X \to K(Y)$, $F_1 : X \to K(Z)$ are upper semicontinuous then their Cartesian product $F_0 \times F_1 : X \to K(Y \times Z)$ is upper semicontinuous.*

Proof. The Tychonoff theorem (see Chapter 0) implies that the multimap $F_0 \times F_1$ has compact values. For an arbitrary point $x \in X$, let $G \supset (F_0 \times F_1)(x)$ be an open subset of $Y \times Z$. From the definition of the product topology in $Y \times Z$ it follows that for every $(y, z) \in (F_0 \times F_1)(x)$ there exist open sets $G_0(y, z) \subset Y$ and $G_1(y, z) \subset Z$ such that $(y, z) \in G_0(y, z) \times G_1(y, z) \subset G$. For each $y \in F_0(x)$ consider the cover $\sum_y = \{G_1(y, \tilde{z}) \mid \tilde{z} \in F_1(x)\}$ of the set $F_1(x)$. Since the set $F_1(x)$ is

compact we can select a finite subcover $\sum'_y = \{G_1(y, \tilde{z}_1), ..., G_1(y, \tilde{z}_k)\}$.
The set $V(y) = \bigcap_{j=1}^{k} G_0(y, \tilde{z}_j)$ is a neighborhood of y in Y and the
set $W_y = \bigcup_{j=1}^{k} G_1(y, \tilde{z}_j)$ is a neighborhood of $F_1(x)$ in Z and moreover
$(V(y) \times W_y) \subset G$. The sets $V(y), y \in F_0(x)$ form an open cover of
the compact set $F_0(x)$. Choose a finite subcover $\{V(y_i)\}_{i=1}^{l}$ and set
$V = \bigcup_{i=1}^{l} V(y_i)$ and $W = \bigcap_{i=1}^{l} W_{y_i}$. Then V is a neighborhood of $F_0(x)$
in Y whereas W is a neighborhood of $F_1(x)$ in Z and $V \times W \subset G$.

Then from Lemma 1.2.12(a) and the upper semicontinuity of the multimaps F_0 and F_1 it follows that there exists a neighborhood $U(x)$ of x such that $(F_0 \times F_1)(U(x)) \subset V \times W \subset G$ proving the upper semicontinuity of $F_0 \times F_1$. ∎

1.3.2   *Algebraic and other operations*

Let X be a topological space, Y a linear topological space.

Definition 1.3.18. Let $F_0, F_1 : X \to P(Y)$ be multimaps. The multimap $F_0 + F_1 : X \to P(Y)$ defined as

$$(F_0 + F_1)(x) = F_0(x) + F_1(x)$$

is called *the sum of the multimaps F_0 and F_1.*

Theorem 1.3.19. *If multimaps $F_0, F_1 : X \to P(Y)$ are lower semicontinuous then their sum $F_0 + F_1 : X \to P(Y)$ is lower semicontinuous.*

Proof. The multimap $F_0 \times F_1 : X \to P(Y \times Y)$ is l.s.c. by Theorem 1.3.15. The single-valued map $f : Y \times Y \to Y$,

$$f(u, v) = u + v$$

is continuous. We have

$$F_0 + F_1 = f \circ (F_0 \times F_1)$$

and conclusion follows from Theorem 1.3.11. ∎

Similar application of Theorems 1.3.17 and 1.3.11 yields the following result.

Theorem 1.3.20. *If multimaps $F_0, F_1 : X \to K(Y)$ are upper semicontinuous then their sum $F_0 + F_1 : X \to K(Y)$ is upper semicontinuous.*

Remark 1.3.21. Notice that the assumption of compactness of the values of the multimaps F_0 and F_1 is essential. In fact, it was mentioned already that the multimap F in Example 1.1.8 is not u.s.c. But it may be represented as the sum of the identity map $F_0(x) \equiv \{x\}$ and the constant multimap $F_1(x) = \{(z_1, z_2) \mid (z_1, z_2) \in \mathbb{R}^2,\ z_1 z_2 = 1,\ z_1 > 0, z_2 > 0\}$.

Definition 1.3.22. Let $F : X \to P(Y)$ be a multimap, $f : X \to \mathbb{R}$ a function. The multimap $f \cdot F : X \to P(Y)$,

$$(f \cdot F)(x) = f(x) \cdot F(x)$$

is called *the product of f and F*.

Theorem 1.3.23. *If a multimap $F : X \to P(Y)$ is lower semicontinuous and a function $f : X \to \mathbb{R}$ is continuous then the product $f \cdot F : X \to P(Y)$ is lower semicontinuous.*

Proof. The multimap $f \times F : X \to P(\mathbb{R} \times Y)$ is l.s.c. by Theorem 1.3.15. The map $\varphi : \mathbb{R} \times Y \to Y$,

$$\varphi(r, y) = r \cdot y$$

is continuous. Then the multimap

$$f \cdot F = \varphi \circ (f \times F)$$

is l.s.c. by Theorem 1.3.11. ∎

The following statement can be proved by a similar application of Theorems 1.3.17 and 1.3.11.

Theorem 1.3.24. *If a multimap $F : X \to K(Y)$ is upper semicontinuous and a function $f : X \to \mathbb{R}$ is continuous then the product $f \cdot F : X \to K(Y)$ is upper semicontinuous.*

Definition 1.3.25. Let Y be a linear topological space, $F : X \to P(Y)$ a multimap. The multimap $\overline{co}F : X \to Cv(Y)$,

$$(\overline{co}F)(x) = \overline{co}(F(x))$$

is called *the convex closure of the multimap F*.

Theorem 1.3.26. *Let Y be a Banach space. If a multimap $F : X \to K(Y)$ is u.s.c. (l.s.c.) then the convex closure $\overline{co}F : X \to Kv(Y)$ is u.s.c. (l.s.c.)*

Proof. First, we note that the multimap $\overline{co}F$ has compact values by Mazur's theorem (see Chapter 0). Let the multimap F be u.s.c. Consider a point $x \in X$ and let $\varepsilon > 0$. Then for every $\varepsilon_1, 0 < \varepsilon_1 < \varepsilon$ there exists a neighborhood $U(x)$ of x such that $F(x') \subset F_{\varepsilon_1}(x)$ for all $x' \in U(x)$ (Theorem 1.2.39). But $F_{\varepsilon_1}(x) \subset U_{\varepsilon_1}(\overline{co}F(x))$, hence $coF(x') \subset U_{\varepsilon_1}(\overline{co}F(x))$ since the set $U_{\varepsilon_1}(\overline{co}F(x))$ is convex. Then

$$\overline{co}F(x') \subset U_{\varepsilon_1}(\overline{co}F(x))$$

for each $x' \in U(x)$ proving, by Theorem 1.2.39, the upper semicontinuity of the multimap $\overline{co}F$.

The lower semicontinuity of the multimap $\overline{co}F$ can be proved in a similar way by applying Theorem 1.2.40. ∎

Remark 1.3.27. The property of closedness of a multimap can be lost under the operation of convex closure, as the following example shows.

Example 1.3.28. The multimap $F : \mathbb{R} \to C(\mathbb{R})$,

$$F(x) = \begin{cases} \{0\}, & x \le 0, \\ \{\pm \frac{1}{x}\}, & x > 0 \end{cases}$$

is closed, but its convex closure $\overline{co}F : \mathbb{R} \to Cv(\mathbb{R})$,

$$\overline{co}F(x) = \begin{cases} \{0\}, & x \le 0, \\ \left[-\frac{1}{x}, \frac{1}{x}\right], & x > 0 \end{cases}$$

is not closed.

1.3.3 Theorem of maximum

Theorem of maximum, which is called sometimes *the principle of continuity of optimal solutions* plays an important role in the applications of multivalued maps in the theory of games and mathematical economics (see Chapter 4).

Theorem 1.3.29. *Let X, Y be topological spaces, $\Phi : X \to K(Y)$ a continuous multimap, $f : X \times Y \to \mathbb{R}$ a continuous function. Then the function $\varphi : X \to \mathbb{R}$,*

$$\varphi(x) = \max_{\tilde{y} \in \Phi(x)} f(x, \tilde{y})$$

is continuous and the multimap $F : X \to P(Y)$

$$F(x) = \{y | y \in \Phi(x), f(x, y) = \varphi(x)\}$$

has compact values and is upper semicontinuous.

Remark 1.3.30. The function φ and the multimap F are often called *marginal.*

The proof of Theorem 1.3.29 will be based on the following two assertions.

Lemma 1.3.31. *Let a multimap $\Phi : X \to K(Y)$ be lower semicontinuous, a function $f : X \times Y \to \mathbb{R}$ lower semicontinuous (in the single-valued sense). Then the function $\varphi : X \to \overline{\mathbb{R}}$,*

$$\varphi(x) = \sup_{\tilde{y} \in \Phi(x)} f(x, \tilde{y})$$

is lower semicontinuous.

Proof. Choose a point $x \in X$ and assume at first that $\varphi(x) < +\infty$. Fix $\varepsilon > 0$; then there exists a point $y \in \Phi(x)$ such that $f(x, y) \geq \varphi(x) - \varepsilon$. By the lower semicontinuity of f there exist neighborhoods $U_0(x)$ of x and $V(y)$ of y such that, for each $x' \in U_0(x), y' \in V(y)$ we have

$$f(x', y') > f(x, y) - \varepsilon \geq \varphi(x) - 2\varepsilon.$$

By the lower semicontinuity of the multimap Φ there exists a neighborhood $U_1(x)$ of x such that $x'' \in U_1(x)$ implies $\Phi(x'') \cap V(y) \neq \emptyset$. Further, if $\tilde{x} \in U(x) = U_0(x) \cap U_1(x)$ then there exists $\tilde{y} \in \Phi(x'') \cap V(y)$ and then $f(\tilde{x}, \tilde{y}) > \varphi(x) - 2\varepsilon$, hence $\varphi(\tilde{x}) > \varphi(x) - 2\varepsilon$.

The case $\varphi(x) = +\infty$ can be considered similarly. ∎

Lemma 1.3.32. *Let a multimap $\Phi : X \to K(Y)$ be upper semicontinuous, a function $f : X \times Y \to \mathbb{R}$ be upper semicontinuous (in the single-valued sense). Then the function $\varphi : X \to \mathbb{R}$,*

$$\varphi(x) = \max_{\tilde{y} \in \Phi(x)} f(x, \tilde{y})$$

is upper semicontinuous.

Proof. Fix $\varepsilon > 0$. For each pair $x \in X, y \in \Phi(x)$ there exist neighborhoods $U_y(x), V(y)$ such that $x' \in U_y(x), y' \in V(y)$ implies $f(x', y') < f(x, y) + \varepsilon$. Since the set $\Phi(x)$ is compact, there exist a finite number of points $y_1, ..., y_n$ such that the neighborhoods $V(y_i), 1 \leq i \leq n$ form a cover of $\Phi(x)$. If now $U_0(x) = \bigcap_{i=1}^{n} U_{y_i}(x)$, and $V(\Phi(x)) = \bigcup_{i=1}^{n} V(y_i)$ then from $x'' \in U_0(x), y'' \in V(\Phi(x))$ it follows that

$$f(x'', y'') < \max_{1 \leq i \leq n} f(x, y_i) + \varepsilon \leq \varphi(x) + \varepsilon.$$

Let $U_1(x)$ be a neighborhood of x such that $\Phi(U_1(x)) \subset V(\Phi(x))$. Then $\tilde{x} \in U(x) = U_0(x) \cap U_1(x)$ yields $\Phi(\tilde{x}) \subset V(\Phi(x))$ and for each $\tilde{y} \in \Phi(\tilde{x})$ we have $f(\tilde{x}, \tilde{y}) < \varphi(x) + \varepsilon$ implying $\varphi(\tilde{x}) < \varphi(x) + \varepsilon$.   ■

Proof of Theorem 1.3.29. For every $x \in X$, the set

$$\Gamma(x) = \{y | y \in Y, f(x, y) = \varphi(x)\}$$

is nonempty. From Lemmas 1.3.31 and 1.3.32 it follows that the function φ is continuous but then the multimap $\Gamma : X \to C(Y)$ is closed (see Example 1.2.27). Now, notice that $F = \Phi \cap \Gamma$ and apply Theorem 1.3.3.   ■

1.4 Continuous selections and approximations of multivalued maps

> Beyond each corner, new directions lie in wait.
>
> *—Stanislaw Jerzy Lec*

Let X, Y be topological spaces, $F : X \to P(Y)$ a multimap.

Definition 1.4.1. A single-valued map $f : X \to Y$ is called *a selection of a multimap F* if

$$f(x) \in F(x)$$

for all $x \in X$.

The problem of existence of selections possessing certain properties is very interesting and finds various applications in many branches of contemporary mathematics. The question on the existence of measurable selections will be discussed in Section 1.5, here we will deal with continuous selections.

The following continuous selection theorem is based on the classical theorem of E. Michael [306].

Theorem 1.4.2. *Let X be a paracompact topological space; Y a Banach space. Then every lower semicontinuous multimap $F : X \to Cv(Y)$ admits a continuous selection.*

The proof of this theorem essentially uses the following assertion on existence of an approximate selection. Recall that given a multimap F, by the symbol F_ε, where $\varepsilon > 0$ we denote the ε-enlargement of F (see Definition 1.2.38).

Lemma 1.4.3. *Let X be a paracompact topological space, Y a normed space, $F : X \to Pv(Y)$ a lower semicontinuous multimap. Then for every*

$\varepsilon > 0$ *there exists a continuous single-valued map* $f_\varepsilon : X \to Y$ *such that* $f_\varepsilon(x) \in F_\varepsilon(x)$ *for each* $x \in X$.

Proof. Let $U_y = F_-^{-1}(B_\varepsilon(y))$ for every $y \in Y$. These sets are open by virtue of lower semicontinuity of the multimap F. The collection $\{U_y\}_{y \in Y}$ forms an open cover of the paracompact space X. Let $\{V_{y_j}\}_{j \in J}$ be a subordinated locally finite cover. Consider the corresponding partition of unity $\{p_{y_j}\}_{j \in J}$. Define a continuous map $f_\varepsilon : X \to Y$ in the following way:

$$f_\varepsilon(x) = \sum_{j \in J} p_{y_j}(x) \, y_j.$$

It is easy to see that the map f_ε is the desirable one. ∎

Proof of Theorem 1.4.2. Construct by induction the sequence of continuous maps $\{f_k\}_{k=1}^{\infty}$,

$$f_k : X \to Y,$$

satisfying the conditions:

(1) $\|f_{k+1}(x) - f_k(x)\| < \frac{1}{2^{k-1}}$;
(2) $f_k(x) \in F_{2^{-k}}(x)$ for every $x \in X$.

The existence of f_1 satisfying condition (2) follows from Lemma 1.4.3. If $f_1, ..., f_k$ are already built then f_{k+1} is constructed in the following way. By the assumption of induction,

$$F^{k+1}(x) = F(x) \cap U_{2^{-k}}(f_k(x)) \neq \emptyset, \ \forall x \in X.$$

From Theorem 1.3.9 it follows that the multimap $F^{k+1} : X \to Pv(Y)$ is lower semicontinuous. Then by using Lemma 1.4.3 we find a continuous map $f_{k+1} : X \to Y$ such that

$$f_{k+1}(x) \in F_{2^{-k-1}}^{k+1}(x)$$

for every $x \in X$. But then for each $x \in X$ we have

$$\|f_{k+1}(x) - f_k(x)\| < \frac{1}{2^k} + \frac{1}{2^{k+1}} < \frac{1}{2^{k-1}},$$

i.e., condition (1) is fulfilled and

$$f_{k+1}(x) \in F_{2^{-k-1}}(x)$$

that means the fulfilment of condition (2) also.

From condition (1) it follows that the sequence $\{f_k\}_{k=1}^{\infty}$ uniformly converges to a continuous function f for which, by virtue of condition (2) and the closedness of values of the multifunction F we get $f(x) \in F(x)$ for every $x \in X$, i.e., f is the desired selection.　　　　　　■

Remark 1.4.4. The complete content of the Michael theorem includes also the assertion that for a T_1 -space X the existence of a continuous selection for each lower semicontinuous multimap $F : X \to Cv(Y)$, where Y is a Banach space is a sufficient condition of the paracompactness of X. The proof of this result can be found in [306], [71].

From the proved theorem the existence of extension for a given selection follows.

Corollary 1.4.5. *Let X be a paracompact space, Y a Banach space, A a closed subset of X. Let $F : X \to Cv(Y)$ be a lower semicontinuous multimap. If a map $g : A \to Y$ is a continuous selection of the multimap F, restricted to A, then there exists a continuous map $f : X \to Y$ such that:*

1) *f is a continuous selection of F;*
2) *$f(x) = g(x)$ for every $x \in A$.*

Proof. Consider the multimap $\widetilde{F} : X \to Cv(Y)$,

$$\tilde{F}(x) = \begin{cases} g(x), & if \ x \in A, \\ F(x), & if \ x \notin A \end{cases}$$

It is easy to verify (**check it!**) that this multimap is also lower semicontinuous and conditions of Theorem 1.4.2 are fulfilled for it. Hence the multimap $\widetilde{F}$ admits a continuous selection f which is the desirable one.　　■

Restrictions established by the conditions of Theorem 1.4.2 are sufficiently severe and the existence of a continuous selection can be guaranteed far from always. First of all, let us mention the importance of the requirement of completeness of the space Y and closedness of values of the multimap F in the conditions of the indicated theorem. Relevant counterexamples can be found in [306].

The following example illustrates the importance of the requirement of convexity of values of the multimap F. The presented multimap will be not only upper semicontinuous, but continuous however it does not admit a continuous selection since its values are non-convex sets.

Example 1.4.6. Let $\overline{B_1}(0) \subset \mathbb{R}^2$ be a closed unit disk on a plane centered at the origin, S its boundary. Consider the multimap $F : \overline{B_1}(0) \to$

$K\left(\overline{B_1}\left(0\right)\right)$, defined in the following way:

$$F\left(x\right) = \begin{cases} S\backslash B_{||x||}\left(\frac{x}{||x||}\right), & x \neq 0, \\ S, & x = 0. \end{cases}$$

It is easy to see that the multimap F is continuous. Assume now that it admits a continuous selection $f : \overline{B_1}\left(0\right) \to \overline{B_1}\left(0\right).$

According to the Brouwer theorem (see Chapter 0) the map f has a fixed point, i.e., there exists a point $x_0 \in \overline{B_1}\left(0\right)$ such that $x_0 = f\left(x_0\right).$

But then $x_0 \in F\left(x_0\right)$ that is impossible by the construction of F.

Nevertheless, the existence of continuous selections may be proved also for certain classes of multimaps with non-convex values. Introduce necessary notions. Let $I = [a, b] \subset \mathbb{R}$ be an interval equipped with the Lebesgue measure; E a Banach space. The symbol $L^1\left(I; E\right)$ denotes the Banach space of equivalence classes of Bochner integrable functions $\varphi : I \to E$ with the norm

$$\| \varphi \| = \int_I \| \varphi\left(s\right) \|_E \, ds.$$

For a measurable subset $m \subset I$, let $\kappa_m : I \to [0, 1]$ denote its characteristic function

$$\kappa_m\left(t\right) = \begin{cases} 1, & t \in m, \\ 0, & t \notin m. \end{cases}$$

Definition 1.4.7. A set $M \subset L^1\left(I; E\right)$ is called *decomposable* if for each $\varphi, \psi \in M$ and a measurable subset $m \subset I$ the function

$$\kappa_m \cdot \varphi + \kappa_{I \backslash m} \cdot \psi$$

belongs to M.

The collection of all nonempty closed decomposable subsets of the space $L^1\left(I; E\right)$ will be denoted by $D\left(L^1\left(I; E\right)\right).$

The following analogue of the Michael selection theorem which is due to Fryszkowski–Bressan–Colombo holds true. (see [142], [143], [83]).

Theorem 1.4.8. *Let X be a separable metric space. Then each lower semicontinuous multimap*

$$F : X \to D\left(L^1\left(I; E\right)\right)$$

admits a continuous selection.

The proof of this assertion can be found in [116], [143], [210].

Let us return to examples of non-existence of continuous selections. One more, a very simple example, is of a special interest for us. It shows that it is impossible to guarantee the existence of a continuous selection even for a multimap with compact convex values if we are leaving the class of lower semicontinuous multimaps.

Example 1.4.9. Consider the multimap
$F : [0,1] \to Kv([0,1])$

$$F(x) = \begin{cases} \{0\}, & 0 \le x \le \frac{1}{2}, \\ [0,1], & x = \frac{1}{2}, \\ \{1\}, & \frac{1}{2} < x \le 1. \end{cases}$$

It is evident that the multimap F is upper semicontinuous (and closed) but it can not have a continuous selection.

So, we see that the collection of multivalued maps admitting single-valued continuous selections is sufficiently narrow: in particular, it does not include upper semicontinuous multimaps, very important for applications. However there exists the possibility of studying the upper semicontinuous multimaps by means of single-valued maps. For a sufficiently wide class of such multimaps their *single-valued continuous approximations* may be constructed.

Let $(X, \varrho_X), (Y, \varrho_Y)$ be metric spaces.

Definition 1.4.10. Let $F : X \to P(Y)$ be a multimap. For a given $\varepsilon > 0$, a continuous map $f_\varepsilon : X \to Y$ is called ε-*approximation of the multimap* F if for each $x \in X$ there exists $x' \in X$ such that $\varrho_X(x, x') < \varepsilon$ and

$$f_\varepsilon(x) \in F_\varepsilon(x').$$

It is clear that this notion can be equivalently expressed by the condition

$$f_\varepsilon(x) \in F_\varepsilon(B_\varepsilon(x))$$

for all $x \in X$.

If we introduce the metric ϱ in the Cartesian product $X \times Y$ by means of the equality

$$\varrho((x,y),(x',y')) = \max\{\varrho_X(x,x'), \varrho_Y(y,y')\},$$

we obtain one more, geometrically visual interpretation: the graph Γ_{f_ε} is contained in the ε-neighborhood of the graph Γ_F.

The following assertion on the existence of an ε-approximation holds.

Theorem 1.4.11. *Let (X, ϱ) be a metric space, Y a normed space. For each upper semicontinuous multimap $F : X \to Cv(Y)$ and every $\varepsilon > 0$ there exists a continuous map $f_\varepsilon : X \to Y$ such that:*

(i) for each $x \in X$ there exists $x' \in X$ such that $\varrho(x, x') < \varepsilon$ and
$$f_\varepsilon(x) \cup F(x) \subset F_\varepsilon(x');$$

(ii) $f_\varepsilon(X) \subset coF(X)$.

Proof. Let us fix $\varepsilon > 0$. For each $x \in X$ there exists $\delta(x) \in (0, \varepsilon)$ such that $F\left(B_{\delta(x)}(x)\right) \subset F_\varepsilon(x)$. For $\eta(x) = \frac{1}{4}\delta(x)$ consider the cover $\left\{B_{\eta(x)}(x)\right\}_{x \in X}$ of the space X and let $\{V_j\}_{j \in J}$ be a subordinated locally finite cover and $\{p_j\}_{j \in J}$ the corresponding partition of unity.

Choosing for each index $j \in J$ an arbitrary point $y_j \in F(V_j)$, define the map $f_\varepsilon : X \to Y$ by the equality
$$f_\varepsilon(x) = \sum_{j \in J} p_j(x) y_j.$$

The map f_ε is the desired one. In fact, let $x \in X$ belongs to all members of the family $\{V_j\}_{j=1}^n$ from the cover $\{V_j\}_{j \in J}$. Every V_j, $j = 1, ..., n$ is contained in some ball $B_{\eta(x_j)}(x_j)$, therefore $x \in \bigcap_{j=1}^n B_{\eta(x_j)}(x_j)$. Let k, $1 \le k \le n$, be such that $\eta_k = \max_{1 \le j \le n} \eta(x_j)$. Take $x' = x_k$, then $x_j \in B_{\eta_k}(x)$ and hence $x_j \in B_{2\eta_k}(x')$ for all $j = 1, ..., n$. Then $B_{\eta(x_j)}(x_j) \subset B_{4\eta_k}(x')$, $j = 1, ..., n$.

But then we get
$$y_j \in F(V_j) \subset F\left(B_{\eta(x_j)}(x_j)\right) \subset F\left(B_{4\eta_k}(x')\right) \subset F_\varepsilon(x')$$

for all $j = 1, ..., n$, and since the set $F_\varepsilon(x')$ is convex, we have $f_\varepsilon(x) \in F_\varepsilon(x')$. Since $x \in V_j, j = 1, ..., n$, we obtain also that $F(x) \subset F_\varepsilon(x')$. $\blacksquare$

Definition 1.4.12. A single-valued ε-approximation satisfying condition (ii) of Theorem 1.4.11, will be called *regular*.

It is easy to see that the restriction of an ε-approximation of a multimap F to a subset of a space X is not, generally speaking, an ε-approximation of the restriction of F to this subset. Nevertheless, the following assertion holds true.

Lemma 1.4.13. *Let X, Y be metric spaces; $F : X \to P(Y)$ a upper semicontinuous multimap; $X_1 \subset X$ a compact subset. Then for every $\varepsilon > 0$*

there exists $\delta > 0$ such that if $f : X \to Y$ is a δ'-approximation of F, $0 < \delta' \leq \delta$, then $f\mid_{X_1}$ is an ε-approximation of $F\mid_{X_1}$.

Proof. Suppose the contrary. Then there exist a number $\varepsilon_0 > 0$, a sequence of δ_n-approximations $f_{\delta_n} : X \to Y$, $(\delta_n \to 0)$ of the multimap F and a sequence of points $\{x_n\} \subset X_1$ such that

$$(x_n, f_{\delta_n}(x_n)) \notin U_{\varepsilon_0}\left(\Gamma_{F\mid_{X_1}}\right), \quad n = 1, 2, \ldots$$

By virtue of compactness of the set X_1, we may assume, without loss of generality, that $x_n \to x_0 \in X_1$. Let η, $0 < \eta \leq 2\varepsilon_0$ be such that $F(B_\eta(x_0)) \subset F_{\varepsilon_0/2}(x_0)$.

Let us find such number N_0 that $\delta_n < \min\{\varepsilon_0/2, \eta/2\}$ and $\varrho(x_n, x_0) < \eta/2$ for all $n \geq N_0$. But then for the same n we have

$$f_{\delta_n}(x_n) \in F_{\delta_n}(B_{\delta_n}(x_n)) \subset F_{\delta_n}(B_\eta(x_0)) \subset$$

$$\subset U_{\delta_n}(F_{\varepsilon_0/2}(x_0)) \subset F_{\varepsilon_0}(x_0).$$

Since also $\varrho(x_n, x_0) < \varepsilon_0$, we get the contradiction. ∎

In conclusion, let us mention that Theorem 1.4.11 also has a "decomposable" analogue. In fact, the following assertion holds true (see [83]).

Theorem 1.4.14. *Let X be a metric space; a multimap $F : X \to D\left(L^1(I; E)\right)$ be upper semicontinuous. Then for every $\varepsilon > 0$, the multimap F has an ε-approximation $f_\varepsilon : X \to L^1(I; E)$ such that $f_\varepsilon(X) \subset decF(X)$, where $decF(X)$ denotes the decomposable hull of $F(X)$, i.e., the least decomposable set containing $F(X)$.*

Concerning the existence of single-valued ε-approximations for other classes of nonconvex-valued multimaps see Section "Bibliographic notes and additions".

1.5 Measurable multivalued functions and the superposition multioperator

> If you do not know his language, you will never understand a foreigner's silence.
>
> —*Stanislaw Jerzy Lec*

1.5.1   *Measurable multifunctions and a multivalued integral*

The notion of a measurable multivalued function (multifunction) is a natural extension of a classical one. It turns out to be very useful in numerous applications in analysis, theory of differential equations and inclusions, optimization theory and other branches of contemporary mathematics. In this section we will describe some properties of measurable multifunctions. We do not strive for the maximal generality in our presentation, giving our definitions and assertions in a form convenient for subsequent applications. In particular, we will restrict ourselves to the case of multifunctions defined on an interval of the real line equipped with the Lebesgue measure and taking its values in the collection of compact subsets of a Banach space.

So, let $I \subset \mathbb{R}$ be a compact interval, μ the Lebesgue measure on I and E a Banach space.

Definition 1.5.1. A multifunction $F : I \to K(E)$ is called *measurable* if for every open subset $V \subset E$ the set $F_+^1(V)$ is measurable.

It is clear that an equivalent definition is the condition of measurability of the complete preimage $F_-^{-1}(W)$ of every closed subset $W \subset E$. The following statement gives two more equivalent definitions of the measurability of a multifunction.

Lemma 1.5.2. *A multifunction $F : I \to K(E)$ is measurable if and only if one of the following conditions holds: a) for every closed set $W \subset E$ the small preimage $F_+^{-1}(V)$ is measurable; b) for every open set $V \subset E$ the complete preimage $F_-^{-1}(V)$ is measurable.*

Proof. 1) Let a multifunction F be measurable, $W \subset E$ a closed set. From Lemma 1.2.3(e) it follows that for an arbitrary sequence $\{\varepsilon_n\}_{n=1}^{\infty}, \varepsilon_n > 0, \varepsilon_n \to 0$ we have

$$F_+^{-1}(W) = \bigcap_{n=1}^{\infty} F_+^{-1}(U_{\varepsilon_n}(W))$$

and hence the set $F_+^{-1}(W)$ is measurable.

2) Conversely, if $F_+^{-1}(W)$ is a measurable set for every closed $W \subset E$ then for each open set $V \subset E$ the set $F_-^{-1}(V)$ is measurable. Let $W \subset E$ be a closed set. It is easy to verify (**do it!**) that under the compactness of values of F, for an arbitrary sequence $\{\varepsilon_n\}_{n=1}^{\infty}, \varepsilon_n > 0, \varepsilon_n \to 0$ we have

$$F_-^{-1}(W) = \bigcap_{n=1}^{\infty} F_-^{-1}(U_{\varepsilon_n}(W))$$

and hence $F_-^{-1}(W)$ is measurable.   ∎

Notice that in the case when the space E is separable, we can obtain
the equivalent definitions of measurability for a multifunction assuming the
measurability of small (or complete) preimages only for open (or closed)
balls in E of a rational radius centered at the points of a countable dense
subset. **Verify it** by using the fact that such open balls form a base of
topology of E.

From Definition 1.5.1 and Lemma 1.5.2 it follows that each u.s.c. or
l.s.c. multifunction $F : I \to K(E)$ is measurable.

To describe additional properties of measurable multifunctions we need
the following notions.

Definition 1.5.3. A function $f : I \to E$ is called *a measurable selection* of
a multifunction $F : I \to K(E)$ if f is measurable and

$$f(t) \in F(t)$$

for μ-a.e. $t \in I$.

The set of all measurable selections of F will be denoted as S_F.

Definition 1.5.4. A countable family $\{f_n\}_{n=1}^{\infty} \subset S_F$ is said to be *a Castaing representation* for F if

$$\overline{\bigcup_{n=1}^{\infty} f(t)} = F(t)$$

for μ-a.e. $t \in I$.

A multifunction $F : I \to K(E)$ is called *a step multifunction* if there
exists a partition of I into a finite family of disjoint measurable subsets
$\{I_j\}$, $\bigcup_j I_j = I$ such that F is constant on each I_j.

Definition 1.5.5. A multifunction $F : I \to K(E)$ is called *strongly measurable* if there exists a sequence $\{F_n\}_{n=1}^{\infty}$ of step multifunctions such that

$$h(F_n(t), F(t)) \to 0$$

as $n \to \infty$ for μ-a.e. $t \in I$, where h is the Hausdorff metric on $K(E)$ (see
Definition 1.2.42).

It is known that in the same way we can define the concept of a strongly
measurable function and, hence, of a strongly measurable selection. Notice
that, in general, a measurable multifunction is not a strongly measurable

one (see, for example, [116]). But for compact-valued multifunctions acting into a separable Banach space these notions coincide. This becomes clear from the following proposition describing the main properties of measurable multifunctions.

Theorem 1.5.6. *Let E be a separable Banach space. For a multifunction $F : I \to K(E)$ the following conditions are equivalent:*

(a) *F is measurable;*

(b) *for every countable dense subset $\{x_n\}_{n=1}^{\infty}$ of E the functions $\{\varphi_n\}_{n=1}^{\infty}$, $\varphi_n : I \to \mathbb{R}$*

$$\varphi_n(t) = \varrho(x_n, F(t)),$$

are measurable (here ϱ is the metric in E generated by the norm);

(c) *F has a Castaing representation;*

(d) *F is strongly measurable;*

(e) *F is measurable as a single-valued function from I into the metric space $(K(E), h)$;*

(f) *F satisfies the Lusin property: for every $\delta > 0$ there exists a closed subset $I_\delta \subset I$ such that $\mu(I \backslash I_\delta) \leq \delta$ and the restriction of F to I_δ is continuous.*

Proof. 1) $(a) \Leftrightarrow (b)$. Notice that the collection of balls $B_r(x_n)$ of rational radius r forms a countable base of the topology of the space E. Then condition (a) is equivalent to the measurability of the preimage $F_-^{-1}(B_r(x_n))$ of every such ball. But this preimage coincides with the Lebesgue set

$$\Delta_{x_n}(r) = \{t | t \in I, \varphi_n(t) < r\}.$$

2) $(b) \Rightarrow (c)$. Let $Q = \{x_n\}_{n=1}^{\infty}$ be a countable dense subset of E. Define the sequence of functions $\{\psi_k\}_{k=1}^{\infty}$, $\psi_k : I \to E$ by the following inductive process. Set

$$\psi_1(t) = x_i$$

if for a given $t \in I$, i is the minimal index such that

$$\varphi_i(t) = \varrho(x_i, F(t)) \leq \frac{1}{2}.$$

If ψ_k is constructed then ψ_{k+1} is defined as

$$\psi_{k+1}(t) = x_i$$

if for a given t, i is the minimal number such that

$$\varphi_i(t) \leq \frac{1}{2^{k+1}}$$

and

$$\varrho(\psi_k(t), x_i) \leq \frac{1}{2^{k-1}}.$$

Notice that the functions ψ_k are measurable. In fact, each of these functions has at most countable number of values and the Lebesgue sets of the functions φ_i,

$$\bar{\Delta}_i(\alpha) = \{t | t \in I, \varphi_i(t) \leq \alpha\}$$

are measurable. Then the function ψ_1 is measurable, since

$$\{t | t \in I, \psi_1(t) = x_i\} = \bar{\Delta}_i\left(\frac{1}{2}\right) \setminus \bigcup_{p<i} \bar{\Delta}_p\left(\frac{1}{2}\right)$$

is a measurable set. Further, if we suppose that the function ψ_k, $k \geq 1$ is measurable then

$$\{t | t \in I, \psi_{k+1}(t) = r_i\} = \left\{ \bar{\Delta}_i\left(\frac{1}{2^{k+1}}\right) \cap \left[t | t \in I, \varrho(\psi_k(t), x_i) \leq \frac{1}{2^{k-1}}\right] \right\} \setminus$$

$$\setminus \left\{ \bigcup_{p<i} \left[\bar{\Delta}_p\left(\frac{1}{2^{k+1}}\right) \cap \left[t | t \in I, \varrho(\psi_k(t), x_p) \leq \frac{1}{2^{k-1}}\right] \right] \right\}$$

that yields the measurability of ψ_{k+1}.

For every $t \in I$ we have

$$\varrho(\psi_k(t), \psi_{k+1}(t)) \leq \frac{1}{2^{k-1}},$$

so the sequence $\{\psi_k\}_{k=1}^{\infty}$ uniformly converges to a measurable function $\psi_Q : I \to E$.

From the relations

$$\varrho(\psi_k(t), F(t)) \leq \frac{1}{2^{k+1}}$$

it follows that ψ_Q is a measurable selection of F.

Therefore we can assign a measurable selection ψ_Q to every countable dense subset $Q \subset E$. Notice that if for any $t \in I$ and $x \in F(t)$ we have

$$\varrho(x, x_1) \leq \frac{1}{2^k},$$

then $\psi_k(t) = x_1$ and

$$\varrho(x, \psi_Q(t)) \le \frac{1}{2^{k-2}} + \frac{1}{2^k}.$$

Now let $\{Q_m\}_{m=1}^{\infty}$ be a sequence of countable dense subsets of E constructed in the following way. If $Q_0 = \{x_1, ..., x_i, ...\}$ then $Q_m = \{x_{m_1}, ..., x_{m_l}, ...\}$, where $x_{m_r} = x_{m+r}$. Then the sequence of functions $\{f_m\}_{m=1}^{\infty}$,

$$f_m = \psi_{Q_m}$$

forms a Castaing representation of F.

In fact, if $t \in I, x \in F(t)$ and an integer $k > 0$ are given then we can find a number m such that

$$\varrho(x_{m_1}, x) \le \frac{1}{2^k}$$

and hence

$$\varrho(x, \psi_{Q_m}(t)) \le \frac{1}{2^{k-2}} + \frac{1}{2^k}.$$

3) $(c) \Rightarrow (b)$. Let $x \in E$ be any point and $\{f_n\}_{n=1}^{\infty}$ a Castaing representation for F. Then

$$\varphi_x(t) = \varrho(x, F(t)) = \inf_n \| x - f_n(t) \|,$$

so the function φ_x is measurable.

4) $(d) \Leftrightarrow (e) \Leftrightarrow (f)$. It is known that the metric space $(K(E), h)$ is separable (see, for example [135]). Then the indicated equivalences are the direct consequences of the analogous properties of single-valued measurable functions (see, for example [384], Theorem 23_2 and Corollary of Theorem 33, Chapter 4).

5) $(f) \Rightarrow (a)$. Let $W \subset E$ be a closed set. For arbitrary $\delta > 0$ take a closed subset $I_\delta \subset I$ such that $\mu(I \setminus I_\delta) \le \delta$ and the restriction of F to I_δ is continuous. Then the complete preimage $F_-^{-1}(W)$ consists of the closed set $F_-^{-1}(W) \cap I_\delta$ and the set $F_-^{-1}(W) \cap (I \setminus I_\delta)$ whose outer measure is less or equal δ and therefore the set $F_-^{-1}(W)$ is measurable.

6) $(c) \Rightarrow (e)$. Let $M \in K(E)$ and $\{f_n\}_{n=1}^{\infty}$ be a Castaing representation for F. Then

$$h(M, F(t)) = \max \left\{ \sup_n \varrho(f_n(t), M), \sup_{x \in M} \inf_n \| x - f_n(t) \| \right\},$$

so the function $t \to h(M, F(t)), t \in I$ is measurable. ∎

Remark 1.5.7. Notice that the notion of measurability may be introduced also for a closed-valued multifunction $F : I \to C(E)$. From the proof of Theorem 1.5.6 it is clear that in this case conditions (a), (b) and (c) are equivalent too (see [97]).

Corollary 1.5.8. *Let E be a separable Banach space.*

(a) *Let $\{F_j\}_{j \in J}$ be an at most countable family of measurable multi-functions $F_j : I \to K(E)$ such that $\bigcap_{j \in J} F_j(t) \neq \emptyset$ for all $t \in I$. Then the intersection $\bigcap_{j \in J} F_j(t) : I \to K(E)$ is the measurable multifunction;*

(b) *If multifunctions $F_0, F_1 : I \to K(E)$ are measurable then their Cartesian product $F_0 \times F_1 : I \to K(E \times E)$ and the sum $F_0 + F_1 : I \to K(E)$ are measurable multifunctions;*

(c) *If a multifunction $F : I \to K(E)$ is measurable and $f : I \to \mathbb{R}$ is a measurable function then their product $f \cdot F : I \to K(E)$ is a measurable multifunction;*

(d) *If a multifunction $F : I \to K(E)$ is measurable then its convex closure $\overline{co}F : I \to Kv(E)$ is a measurable multifunction.*

Proof. All these statements can be deduced from the Lusin property for measurable multifunctions (Theorem 1.5.6(f)) and the assertions on the continuity of the results of corresponding operations on multimaps (see Section 1.3). ∎

Corollary 1.5.9. *Let E be a separable Banach space. If multifunctions $F_0, F_1 : I \to K(E)$ are measurable then the deviation function $\varrho_* : I \to \mathbb{R}_+$,*

$$\varrho_*(t) = \varrho_*(F_0(t), F_1(t))$$

and the Hausdorff function $h : I \to \mathbb{R}_+$,

$$h(t) = h(F_0(t), F_1(t))$$

are measurable.

Proof. The measurability of the function ϱ_* follows from the equality

$$\varrho_*(t) = \sup_n [\varrho(x_n, F_1(t)) - \varrho(x_n, F_0(t))],$$

where $\{x_n\}_{n=1}^{\infty}$ is a countable dense subset of E and from Theorem 1.5.6(b). The relation

$$h(t) = h(F_0(t), F_1(t)) = \max\{\varrho_*(F_0(t), F_1(t)), \varrho_*(F_1(t), F_0(t))\}$$

implies the measurability of the function h. ∎

Lemma 1.5.10. *Let E be a Banach space. If $\{F_j\}_{j \in J}$, $F_j : I \to K(E)$ is an at most countable family of measurable multifunctions and there exists a multifunction $\Phi : I \to K(E)$ such that*

$$\bigcup_{j \in J} F_j(t) \subset \Phi(t)$$

for μ-a.e. $t \in I$ then the multifunction $\overline{\bigcup_{j \in J} F_j} : I \to K(E),$

$$\overline{\bigcup_{j \in J} F_j}(t) = \overline{\bigcup_{j \in J} F_j(t)}$$

is measurable.

Proof. Let $V \subset E$ be an open set. Then

$$\left(\overline{\bigcup_{j \in J} F_j}\right)^{-1}_{-}(V) = \left(\bigcup_{j \in J} F_j\right)^{-1}_{-}(V) = \bigcup_{j \in J} (F_j)^{-1}_{-}(V)$$

is a measurable set and we can apply Lemma 1.5.2(b). ∎

As we mentioned already, in an arbitrary (not separable) Banach space E a measurable multifunction is not strongly measurable in the general case. At the same time the following assertion holds true.

Theorem 1.5.11. *Let E be a Banach space; $F : I \to K(E)$ a strongly measurable multifunction. Then F is measurable and admits a Castaing representation consisting of strongly measurable functions.*

Proof. Let $\{F_n\}_{n=1}^{\infty}$ be a sequence of step multifunctions approximating the multifunction F. Then for μ-a.e. $t \in I$ the sets $F(t)$ are contained in the separable Banach space

$$E' = \overline{sp} \bigcup_{n=1}^{\infty} F_n(I),$$

where $\overline{sp}$ denotes the closure of the linear hull, and we may apply Theorem 1.5.6. ∎

Consider now the notion of *the multivalued integral*. Let E be a Banach space and $F : I \to P(E)$ a multifunction.

By the symbol S^1_F we will denote the set of all Bochner integrable selections of the multifunction F, i.e.,

$$S^1_F = \left\{ f \mid f \in L^1(I; E), \ f(t) \in F(t) \ \text{for} \ \mu - \text{a.e.} \ t \in I \right\}.$$

If $S_F^1 \neq \emptyset$ then the multifunction F is called *integrable* and

$$\int_{\mathcal{T}} F(s)\,ds := \left\{ \int_{\mathcal{T}} f(s)\,ds \mid f \in S_F^1 \right\}$$

for any measurable subset $\mathcal{T} \subset I$. Clearly if a multifunction $F : I \to K(E)$ is strongly measurable and *integrably bounded,* i.e., there exists an integrable function $\nu \in L_+^1(I)$ such that

$$\|F(t)\| := \max\{\|y\| \mid y \in F(t)\} \leq \nu(t)$$

for μ-a.e. $t \in I$ then F is integrable.

A multivalued integral possesses many interesting properties. We indicate only a few of them connected with the notion of convexity.

Theorem 1.5.12. *Let E be a separable Banach space; $F : I \to K(E)$ a measurable integrable multifunction. Then*

(a) $\overline{\int_I \overline{co}F(s)\,ds} = \overline{co} \int_I F(s)\,ds$;

(b) *the set $\int_I F(s)\,ds$ is convex, moreover, if the space E is finite-dimensional, then the integral $\int_I F(s)\,ds$ is also convex*

(c) *if F is integrably bounded then*

$$\int_I \overline{co}F(s)\,ds = \overline{\int_I F(s)\,ds};$$

(d) *if E is reflexive, F is integrably bounded and has convex values then the integral $\int_I F(s)\,ds$ is closed;*

(e) *if F is constant: $F(t) \equiv A \in Kv(E)$ then*

$$\int_I F(s)\,ds = A\mu(I).$$

The proofs of properties (a)–(d) can be found, for example, in [27], [210]. Property (e) follows from the evident relation $A\mu(I) \subset \int_I F(s)\,ds$ and from the fact that, by the convexity of the set A, the integral sums defining the integral $\frac{1}{\mu(I)} \int_I f(s)\,ds$ for an arbitrary integrable selection f of F, are contained in A.

1.5.2   *The Carathéodory conditions and the Filippov implicit function lemma*

Let E_0, E be Banach spaces; as in the previous section, by I we denote a compact interval endowed with the Lebesgue measure μ.

Definition 1.5.13. A multimap $F : I \times E_0 \to K(E)$ satisfies *upper [lower] Carathéodory conditions* if:

(a) for every $x \in E_0$ the multifunction

$$F(\cdot, x) : I \to K(E)$$

is measurable;

(b) for almost every $t \in I$ the multimap

$$F(t, \cdot) : E_0 \to K(E)$$

is upper semicontinuous [respectively, lower semicontinuous].

If a multimap F satisfies both upper and lower Carathéodory conditions we will say that it satisfies *the Carathéodory conditions.*

In other words, a multimap F satisfies the Carathéodory conditions if it obeys (a) and condition

(b') for almost every $t \in I$ the multimap

$$F(t, \cdot) : E_0 \to K(E)$$

is continuous.

Consider some properties of multimaps satisfying these conditions.

At first notice that if the spaces E_0, E are separable then from Theorem 1.2.44 and Theorem 1.5.6(e) it follows that a multimap $F : I \times E_0 \to K(E)$ satisfying Carathéodory conditions may be considered as a single-valued map into the separable metric space $(K(E), h)$ satisfying "single-valued" Carathéodory conditions. So, we can apply to F the corresponding result for single-valued maps (see, for example [210], Theorem 7.11) and to deduce from it the following *Scorza Dragoni property* being a parametric version of the Lusin property.

Theorem 1.5.14. *Let E_0, E be separable Banach spaces and a multimap $F : I \times E_0 \to K(E)$ satisfies the Carathéodory conditions. Then for every $\delta > 0$ there exists a closed subset $I_\delta \subset I$ such that $\mu(I \setminus I_\delta) \leq \delta$, and the restriction of F to $I_\delta \times E_0$ is continuous.*

Notice that the analogous property for multimaps satisfying the upper or lower Carathéodory conditions does not hold (see Example 1.5.20 (a) and (b) below).

By using the Scorza Dragoni property we can prove the following theorem which is a generalization of the result being very important in the control theory. In its initial form it was proved by A.F. Filippov [137].

In the literature this statement is known as *the Filippov implicit function lemma*

Theorem 1.5.15. *Let E_0, E be separable Banach spaces and a multimap $F : I \times E_0 \to K(E)$ satisfy the Carathéodory conditions. Suppose that $U : I \to K(E_0)$ is a measurable multifunction and $g : I \to E$ is a measurable function such that*

$$g(t) \in F(t, U(t))$$

for μ-a.e. $t \in I$. Then there exists a measurable selection $u \in S_U$ such that

$$g(t) \in F(t, u(t))$$

for μ-a.e. $t \in I$.

Before proving this theorem we state the following proposition which can be easily verified (**do it!**).

Lemma 1.5.16. *Let $F : I \times E_0 \to C(E)$ be a closed multimap, $g : I \to E$ a continuous function. Suppose that for every $t \in I$ there exists $x \in E_0$ such that*

$$g(t) \in F(t, x).$$

Then the multifunction $\Gamma : I \to C(E_0)$,

$$\Gamma(t) = \{x | x \in E_0, g(t) \in F(t, x)\}$$

is closed.

Proof of Theorem 1.5.15. Without loss of generality we can assume that the multimap F is continuous in the second argument and $g(t) \in F(t, U(t))$ for all $t \in I$. Consider the multifunction $\Gamma : I \to C(E_0)$,

$$\Gamma(t) = \{x | x \in E_0, g(t) \in F(t, x)\}.$$

It easy to see that the multifunction $\Phi = \Gamma \cap U : I \to K(E_0)$ is well defined and its measurable selection is the desired function $u : I \to E_0$. Hence, to prove the theorem, it is sufficient to demonstrate the measurability of the multifunction Φ.

To this aim, fix $\delta > 0$. By Theorem 1.5.14 there exists a closed subset $I_1 \subset I$ such that $\mu(I \setminus I_1) \leq \frac{\delta}{2}$ and the restriction of F to $I_1 \times E_0$ is continuous. The application of the Lusin property to the function g and the multifunction U gives the existence of a closed subset $I_2 \subset I$ such that $\mu(I \setminus I_2) \leq \frac{\delta}{2}$ and the restrictions of g and U to I_2 are continuous.

If $I_\delta = I_1 \cap I_2$ then $\mu (I \setminus I_\delta) \le \delta$ and from Theorem 1.2.29 it follows that the restriction of F to $I_\delta \times E_0$ is closed. Then, by Lemma 1.5.16 the multifunction Γ is closed on I_δ. Theorem 1.3.3 yields that the multimap Φ is upper semicontinuous on I_δ, from where, by Theorem 1.5.6(f) the measurability of Φ on I follows. ∎

Remark 1.5.17. Fairly often a more particular case of the Filippov lemma is considered: the multimap F is supposed to be single-valued. In this case, the statement means the existence of a measurable selection $u : I \to E_0$ of the multifunction U such that

$$g(t) = f(t, u(t))$$

for μ-a.e. $t \in I$.

1.5.3 *The superposition multioperator*

Every multimap $F : I \times E_0 \to K(E)$ generates a correspondence assigning to every multifunction $Q : I \to P(E_0)$ the multifunction $\Phi : I \to P(E)$ defined by the formula

$$\Phi(t) = F(t, Q(t)).$$

We will consider the properties of this operator connected with its actions on measurable functions and multifunctions.

First of all, let us indicate the following property of *superpositional measurability*.

Theorem 1.5.18. *If E_0, E are separable Banach spaces and a multimap $F : I \times E_0 \to K(E)$ satisfies the Carathéodory conditions then for every measurable multifunction $Q : I \to K(E_0)$ the multifunction Φ is measurable.*

Proof. Without loss of generality we can assume that the multimap F is continuous in x for all $t \in I$. Then the multifunction Φ has compact values by Theorem 1.2.35.

Fix $\delta > 0$ and choose a closed set $I_\delta \subset I$ such that $\mu(I \setminus I_\delta) \le \delta$ and the restrictions of Q to I_δ and F to $I_\delta \times E_0$ are continuous (Theorems 1.5.6(f) and 1.5.14). If $i : I_\delta \to I_\delta$ is the identity map then from Theorems 1.3.15 and 1.3.17 it follows that the multifunction $i \times Q : I_\delta \to K(I_\delta \times E_0)$ is continuous. Then the multifunction $\Phi = F \circ (i \times Q)$ is also continuous on I_δ (Theorem 1.3.11). The measurability of Φ follows now from Theorem 1.5.6(f). ∎

Taking into account that upper or lower semicontinuous functions are measurable and following the similar line of reasonings we can verify (**do it!**) the following sufficient conditions for superpositional measurability.

Theorem 1.5.19. *If E_0, E are separable Banach spaces and a multimap $F : I \times E_0 \to K(E)$ is upper or lower semicontinuous then it is superpositionally measurable.*

It is essential that the property of superpositional measurability disappears if we weaken the Carathéodory conditions to upper or lower ones. Consider the following examples (see [324]).

Example 1.5.20. Let $D \subset [0,1]$ be an arbitrary non-measurable set.

a) Let the multimap $F : [0,1] \times \mathbb{R} \to Kv(\mathbb{R})$ be defined in the following way:

$$F(t,x) = \begin{cases} [0,1], & \text{if } x = t \quad \text{ and } t \in [0,1] \setminus D; \\ [0,1], & \text{if } x = t+1 \text{ and } t \in D; \\ \{1\} & \text{in remaining cases.} \end{cases}$$

It is easy to see that the multimap F satisfies the upper Carathéodory conditions, but does not obey the Carathéodory conditions.

If now $i : [0,1] \to \mathbb{R}$, $i(t) = t$ is the inclusion function then the multi-function $\Phi : [0,1] \to K(\mathbb{R})$,

$$\Phi(t) = F(t, i(t)) = \begin{cases} [0,1], & t \in [0,1] \setminus D; \\ \{1\}, & t \in D, \end{cases}$$

is not measurable.

b) Let the multimap $F : [0,1] \times \mathbb{R} \to Kv(\mathbb{R})$ be defined in the following way:

$$F(t,x) = \begin{cases} \{0\}, & \text{if } x = t \text{ and } t \in [0,1] \setminus D; \\ \{1\}, & \text{if } x = t \text{ and } t \in D; \\ [0,1] & \text{in remaining cases.} \end{cases}$$

This multimap satisfies only lower Carathéodory conditions. By using again the inclusion function $i : [0,1] \to \mathbb{R}$ we get the multifunction

$$\Phi(t) = F(t, i(t)) = \begin{cases} \{0\}, & t \in [0,1] \setminus D; \\ \{1\}, & t \in D, \end{cases}$$

which is not measurable.

This example demonstrates also that the Scorza Dragoni property (see Theorem 1.5.14) does not hold for multimaps satisfying upper or lower Carathéodory conditions. In fact it is easy to see that every multimap satisfying the Scorza Dragoni property is necessarily superpositionally measurable.

Nevertheless, if F satisfies the upper Carathéodory conditions the resulting multimap Φ possesses some "good" properties and moreover, they are fulfilled in arbitrary (not necessarily separable) Banach spaces. Namely, introduce the following notion. Let E_0, E be Banach spaces.

Definition 1.5.21. A multimap $F : I \times E_0 \to K(E)$ *is superpositionally selectable if for every strongly measurable function* $q : I \to E_0$ *there exists a strongly measurable selection* $f : I \to E$ *of the multifunction* $\Phi : I \to K(E)$,

$$\Phi(t) = F(t, q(t)).$$

Theorem 1.5.22. *Let* E_0, E *be Banach spaces and a multimap* $F : I \times E_0 \to K(E)$ *be such that*

(F1) *for every* $x \in E_0$ *the multifunction* $F(\cdot, x) : I \to K(E)$ *admits a strongly continuous selection;*

(F2) *for* μ*-a.e.* $t \in I$ *the multimap* $F(t, \cdot) : E_0 \to K(E)$ *is upper semicontinuous.*

Then the multimap F *is superpositionally selectable.*

Proof. Approximate a function q by a sequence of step functions $\{q_n\}_{n=1}^{\infty}$, $q_n : I \to E_0$, converging μ-almost everywhere on I to q. From condition (F1) it follows that we can form a sequence of strongly measurable functions $\{f_n\}_{n=1}^{\infty}$, $f_n : I \to E$,

$$f_n(t) \in F(t, q_n(t)) \quad \text{for } \mu - \text{a.e. } t \in I.$$

Redefining these functions on a null measure set, we can assume that the functions of the sequence $\{f_n\}$ take their values in a separable Banach space $E' \subset E$,

$$E' = \overline{sp} \bigcup_{n=1}^{\infty} f_n(I).$$

Now for μ-a.e. $t \in I$ and $m \geq 1$ we can define

$$\Phi_m(t) = \overline{\bigcup_{k=m}^{\infty} f_k(t)}.$$

From condition $(F2)$ and Theorem 1.2.35 it follows that the sets $\Phi_m(t)$ are compact and Theorem 1.5.6(c) implies that $\{\Phi_m\}$, $\Phi_m : I \to K(E')$ is the sequence of measurable multifunctions.

For μ-a.e. $t \in I$ the sets $\Phi_m(t)$, $m \geq 1$ form a decreasing sequence of compact sets and taking into account Corollary 1.5.8(a) we conclude that the multifunction $\widetilde{\Phi} : I \to K(E')$,

$$\widetilde{\Phi}(t) = \bigcap_{m=1}^{\infty} \Phi_m(t) \quad \text{for } \mu - \text{a.e. } t \in I$$

is well defined and measurable. From condition $(F2)$ it follows that

$$\widetilde{\Phi}(t) \subset \Phi(t) = F(t, q(t))$$

for μ-a.e. $t \in I$ and the application of Theorem 1.5.6(c) yields the existence of a function $f \in S_{\widetilde{\Phi}}$ which is strongly measurable since $\widetilde{\Phi}(I) \subset E'$. ∎

Corollary 1.5.23. *Let the spaces E_0, E and the multimap F be as in Theorem 1.5.22. Then for every strongly measurable multifunction $Q : I \to K(E_0)$ there exists a strongly measurable selection $f : I \to E$ of the multifunction Φ, $\Phi(t) = F(t, Q(t))$.*

Proof. From Theorem 1.5.11 we obtain the existence of a strongly measurable selection q of the multifunction Q to which we can apply the previous theorem. ∎

Corollary 1.5.24. *If the spaces E_0, E are separable and the multimap $F : I \times E_0 \to K(E)$ satisfies the upper Carathéodory conditions then F is superpositionally selectable.*

Recall (see Chapter 0) that for a given bounded set A in a normed space the value

$$\|A\| := \sup_{a \in A} \|a\|$$

is called *the norm of A*.

Now we can introduce the following notion.

Definition 1.5.25. Let E_0, E be Banach spaces; a multimap $F : I \times E_0 \to K(E)$ be superpositionally selectable and satisfy the following condition of local integral boundedness:

($F3$) for every $r > 0$ there exists such function $\nu_r \in L^1_+ (I)$ that

$$\| F(t, x) \| \leq \nu_r (t)$$

for all x, $\|x\| \leq r$ and a.e. $t \in I$.

Then the multimap $\mathcal{P}_F : C(I; E_0) \to P(L^1(I; E))$ assigning to every continuous function $q \in C(I; E_0)$ the set of all Bochner integrable selections of the multifunction Φ, $\Phi(t) = F(t, q(t))$ is called *the superposition multioperator* generated by F.

It is clear that the integrability of strongly measurable selections of the multifunction Φ follows from the continuity of q and condition ($F3$). We will study the continuity properties of the superposition multioperator.

At first, consider the case when the multimap F is upper semicontinuous in x.

In the sequel, we will need a few auxiliary assertions.

We start with the following notion.

Definition 1.5.26. Let E be a Banach space. The sequence of functions $\{f_n\}_{n=1}^{\infty} \subset L^1(I; E)$ is called *semicompact* if:

(a) it is *integrably bounded*, i.e., there exists a function $\nu \in L^1_+ (I)$ such that

$$\| f_n(t) \| \leq \nu(t)$$

for a.e. $t \in I$;

(b) the set $\{f_n(t)\}_{n=1}^{\infty}$ is relatively compact in E for a.e. $t \in I$.

Lemma 1.5.27. *A semicompact sequence of functions $\{f_n\}_{n=1}^{\infty}$ is weakly compact, i.e., it contains a weakly convergent subsequence.*

See [235], Proposition 4.2.1.

Lemma 1.5.28 (Mazur). *Let $\{u_n\}_{n=1}^{\infty}$ be a sequence of elements of a normed space weakly convergent to u. Then there exists a double sequence of nonnegative numbers $\{\lambda_{ik}\}_{i=1 \, k=1}^{\infty \quad \infty}$ such that:*

(i) $\sum_{k=i}^{\infty} \lambda_{ik} = 1$ *for all $i = 1, 2, ...$;*

(ii) *for each $i = 1, 2, ...$ there exists a number $k_0 = k_0(i)$ such that $\lambda_{ik} = 0$ for all $k \geq k_0$;*

(iii) the sequence of convex combinations $\{\widetilde{u}_i\}_{i=1}^{\infty}$,

$$\widetilde{u}_i = \sum_{k=i}^{\infty} \lambda_{ik} u_k$$

converges to u with respect to the norm.

See, for example, [129].

Lemma 1.5.29. *Let E be a Banach space. If a sequence of functions $\{f_n\}_{n=1}^{\infty} \subset L^1(I; E)$ converges to a function f with respect to the norm of the space $L^1(I; E)$ then there exists a subsequence $\{f_{n_i}\}$ which converges to f almost everywhere on I.*

See, for example, [384], Chapter IV, Theorem 38.

Under the additional assumption that the generating multivalued map F has convex values the following statement about the closedness of the superposition multioperator holds true.

Theorem 1.5.30. *Let E_0, E be Banach spaces; E_1 a normed space; a multimap $F : I \times E_0 \to Kv(E)$ satisfy conditions $(F1)$–$(F3)$ and a : $L^1(I; E) \to E_1$ a bounded linear operator. Then the composition*

$$a \circ \mathcal{P}_F : C(I; E_0) \to Cv(E_1)$$

is a closed multimap.

Proof. First of all, notice that the convexity of values of the multivalued map $a \circ \mathcal{P}_F$ follows from the convexity of values of F and the linearity of the operator a.

Consider sequences $\{q_n\}_{n=1}^{\infty}$, $q_n \in C(I; E_0)$; $\{z_n\}_{n=1}^{\infty}$, $z_n \in E_1$ such that

$$\lim_{n \to \infty} \|q_n - q\|_C = 0, \quad z_n \in a \circ \mathcal{P}_F(q_n), \quad \lim_{n \to \infty} \|z_n - z\|_{E_1} = 0.$$

Choose a sequence $\{f_n\}_{n=1}^{\infty} \subset L^1(I; E)$, $f_n \in \mathcal{P}_F(q_n)$, $z_n = a(f_n)$. Applying conditions $(F2)$, $(F3)$ and Theorem 1.2.35 we can deduce from the convergence of the sequence $\{q_n\}$ that the sequence $\{f_n\}$ is semicompact and hence, by Lemma 1.5.27 it is weakly compact. Passing to a subsequence, if necessary, we will assume, without loss of generality, that it weakly converges to a function $f \in L^1(I; E)$.

Applying now Lemma 1.5.28 we get the sequence $\left\{\widetilde{f}_i\right\}_{i=1}^{\infty}$, $\widetilde{f}_i \in L^1(I; E)$, $\widetilde{f}_i = \sum_{k=i}^{\infty} \lambda_{ik} f_k$ which converges to f with respect to the norm of the space $L^1(I; E)$. By Lemma 1.5.29, we again can assume without loss of generality that the sequence $\left\{\tilde{f}_i\right\}$ converges to f almost everywhere on I.

From condition $(F2)$ it follows that, given a number $\varepsilon > 0$, for almost every $t \in I$ there exists a positive integer $i_0 = i_0(\varepsilon, t)$ such that

$$F(t, q_i(t)) \subset F_\varepsilon(t, q(t)) \quad \text{for} \quad i \geq i_0.$$

But then also

$$f_i(t) \in F_\varepsilon(t, q(t))$$

for all $i \geq i_0$, and hence, by convexity of the set $F_\varepsilon(t, q(t))$ we get

$$\tilde{f}_i(t) \in F_\varepsilon(t, q(t)) \quad \text{for} \quad i \geq i_0.$$

This implies

$$f(t) \in F(t, q(t)) \quad \text{a.e.} \quad t \in I,$$

i.e., $f \in \mathcal{P}_F(q)$.

On the other side,

$$a\left(\tilde{f}_i\right) = \sum_{k=i}^\infty \lambda_{ik} a(f_k) = \sum_{k=i}^\infty \lambda_{ik} z_k,$$

implying $\lim\limits_{i \to \infty} \left\| a\left(\tilde{f}_i\right) - z \right\|_{E_1} = 0$.

From the continuity of the operator a it follows that $z = a(f)$ yielding $z \in a \circ \mathcal{P}_F(q)$ and it remains to apply Theorem 1.2.24(c). ∎

Corollary 1.5.31. *Assume that a multimap* $F : I \times E_0 \to Kv(E)$ *satisfies conditions* $(F1)$–$(F3)$. *Then the superposition multioperator* $\mathcal{P}_F : C(I; E_0) \to Cv\left(L^1(I; E)\right)$ *is closed.*

Remark 1.5.32. Moreover, from the proof of Theorem 1.5.30 we can see that the superposition multioperator $\mathcal{P}_F$ *is weakly closed* in the following sense. Let sequences $\{q_n\}_{n=1}^\infty$, $q_n \in C(I; E_0)$ and $\{f_n\}_{n=1}^\infty \subset L^1(I; E)$, $f_n \in \mathcal{P}_F(q_n)$ be such that $\lim\limits_{n \to \infty} \|q_n - q\|_C = 0$, $\{f_n\}$ weakly converges to $f \in L^1(I; E)$. Then $f \in \mathcal{P}_F(q)$.

Let $I = [t_0, T]$ and $j : L^1(I; E) \to C(I; E)$ be the operator of integration

$$j(f)(t) = \int_{t_0}^t f(s)\, ds.$$

Definition 1.5.33. The composition

$$j \circ \mathcal{P}_F : C\left(I; E_0\right) \to Cv\left(C\left(I; E\right)\right)$$

is called *the integral multioperator* generated by the multimap F.

Corollary 1.5.34. *Assume that a multioperator $F : I \times E_0 \to Kv\left(E\right)$ satisfies conditions $(F1)$–$(F3)$. Then the integral multioperator $j \circ \mathcal{P}_F : C\left(I; E_0\right) \to Cv\left(C\left(I; E\right)\right)$ is closed.*

Consider now the case when the multimap F is lower semicontinuous in x. Notice that the lower Carathéodory conditions do not provide the existence of the superposition multioperator since, as Example 1.5.20(b) demonstrates, under these conditions not only do not we have the superpositional measurability, but the superpositional selectability as well (**verify it!**).

One of appropriate substitutions of lower Carathéodory conditions in this case provides the hypothesis of almost lower semicontinuity.

Let E_0, E be separable Banach spaces.

Definition 1.5.35. A multimap $F : I \times E_0 \to K\left(E\right)$ is called *almost lower semicontinuous* if

(F_L) there exists a sequence of disjoint closed subsets $\{I_n\}$, $I_n \subset I$ such that: (a) $\mu\left(I \setminus \bigcup_n I_n\right) = 0$ and (b) the restriction of F to each set $J_n = I_n \times E_0$ is lower semicontinuous.

From Theorem 1.5.19 it follows that an almost lower semicontinuous multimap F is superpositionally measurable and hence it generates the superposition multioperator $\mathcal{P}_F$.

Lemma 1.5.29 yields that $\mathcal{P}_F$ has closed values. The next assertion also holds true.

Theorem 1.5.36. *Let a multimap $F : I \times E_0 \to K\left(E\right)$ is almost lower semicontinuous and satisfies condition $(F3)$. Then the superposition multioperator $\mathcal{P}_F : C\left(I; E_0\right) \to C\left(L^1\left(I; E\right)\right)$ is lower semicontinuous.*

For the proof we need two auxiliary statements.

Lemma 1.5.37. *For any function $q_0 \in C\left(I; E_0\right)$ and each $\varepsilon > 0$ there exists $\delta > 0$ such that*

$$F\left(t, q_0\left(t\right)\right) \subset F_\varepsilon\left(t, q\left(t\right)\right) \quad a.e. \ t \in I$$

for every function $q \in C\left(I; E_0\right)$, $\|q - q_0\|_C < \delta$.

Proof. Since the multifunction $F(t, q_0(t))$ is measurable, by using the Lusin property (Theorem 1.5.6(f)) and Theorem 1.2.44 we obtain that for every $\rho > 0$ there exist a closed set $I_\rho \subset I$, $\mu(I \setminus I_\rho) \leq \rho$ and a number $\omega > 0$ such that

$$h\left(F(t_1, q_0(t_1)), F(t_2, q_0(t_2))\right) < \frac{\varepsilon}{2}$$

for all $t_1, t_2 \in I_\rho$, $|t_1 - t_2| < \omega$, where h is the Hausdorff metric.

Since F is almost lower semicontinuous, we can cover a compact set, the graph $\Gamma_{q_0} \subset I \times E_0$ of the function q_0 by a finite number of balls

$$B_i = \{(t, x) \,|\, (t, x) \in I \times E_0 : \; |t_i - t| < \omega_i, \; \|q_0(t_i) - x\|_{E_0} < \omega_i\},$$

$$t_i \in I, \; i = 1, ..., m; \; 0 < \omega_i < \omega$$

so that we have

$$F(t_i, q_0(t_i)) \subset F_{\frac{\varepsilon}{2}}(t, x)$$

for all $(t, x) \in B_i \cap (\bigcup_n J_n)$.

Then we get

$$F(t, q_0(t)) \subset F_{\frac{\varepsilon}{2}}(t_i, q_0(t_i)) \subset F_\varepsilon(t, x)$$

for all $(t, x) \in B_i \cap (\bigcup_n J_n)$, $t \in I_\rho$. From the arbitrary choice of ρ it follows that the last relation is fulfilled for a.e. $t \in I$, $(t, x) \in B_i$.

For finding a desirable number $\delta > 0$ it is sufficient now to choose a δ-neighborhood of the compact set Γ_{q_0} contained in $\bigcup_{i=1}^m B_i$. ∎

Lemma 1.5.38. *If $q \in C(I; E_0)$ and $z : I \to E$ is a measurable function then there exists a function $f \in \mathcal{P}_F(q)$ such that*

$$\|z(t) - f(t)\|_E = \varrho(z(t), F(t, q(t))) \; \text{for a.e. } t \in I,$$

where ϱ is the distance from a point to a set generated by the norm in E.

The proof of this statement is not difficult and it is based on the fact that the intersection of measurable multifunctions is measurable (cf. Corollary 1.5.8(a)). But since we considered only compact-valued measurable multifunctions, we refer the reader to [392], Lemma 2.1.4 or [116], Proposition 3.5(b).

Proof of Theorem 1.5.36. From Lemma 1.5.38 it follows that for every integrable function $z : I \to E$ we have

$$\varrho_{L^1}(z, \mathcal{P}_F(q)) = \int_I \varrho(z(s), F(s, q(s))) ds.$$

Take an arbitrary function $q_0 \in C\left(I; E_0\right)$ and choose $\varepsilon > 0$. According to Lemma 1.5.37, let $\delta > 0$ be such that

$$F\left(t, q_0\left(t\right)\right) \subset F_\varepsilon\left(t, q\left(t\right)\right) \quad \text{a.e. } t \in I$$

for each function $q \in C\left(I; E_0\right)$, $\|q - q_0\|_{E_0} < \delta$.

Now, if we take an arbitrary function $z \in \mathcal{P}_F\left(q_0\right)$ then from the last relations we obtain that

$$\varrho_{L^1}\left(z, \mathcal{P}_F\left(q\right)\right) < \varepsilon \cdot \mu\left(I\right),$$

that means

$$\mathcal{P}_F\left(q_0\right) \subset U_{\varepsilon \cdot \mu(I)}\left(\mathcal{P}_F\left(q\right)\right).$$

To get the conclusion about the lower semicontinuity of the superposition multioperator $\mathcal{P}_F$ it remains to apply the sufficiency part of Theorem 1.2.40 (as it was mentioned already, the compactness of values of a multimap is not essential there). ∎

Chapter 2

Fixed points and topological degree

The problem on a solvability of inclusions of various types forms one of the most important parts of contemporary multivalued analysis. In many cases it takes the form of a fixed point problem.

For some sets $X \subset Y$, let $F : X \to P(Y)$ be a multimap. A point $x \in X$ is called *a fixed point of the multimap F if $x \in F(x)$. The set of all fixed points* of F will be denoted by $FixF$.

It is clear that the concept of a fixed point of a multimap is the direct extension of the notion of a fixed point for a usual, single-valued map. The important role which various fixed point principles play in theorems of functional analysis, topology, theory of differential equations and other branches of mathematics is well known. The idea of a fixed point of a multimap also naturally arises in many problems and turns out to be very useful while consideration of numerous applied questions. In the sequel we will be convinced in it by examples from theory of linear operators, theory of differential inclusions and control systems, theory of generalized dynamical systems, as well as from theory of games and mathematical economics.

Similarly to the case of single-valued maps, there exist different approaches to the study of fixed points of multimaps. Describe some of them.

2.1 Fixed points of contractive multimaps

2.1.1   *The Nadler theorem*

The Banach contraction principle is one of the most known and used in applications fixed point results. Let us present the Nadler theorem which is its extension to the multivalued case.

Let (X, ϱ) be a complete metric space; $C(X)$ as earlier, denote the collection of all nonempty closed subsets of X; h be an extended Hausdorff metric in $C(X)$ generated by ϱ (see Definition 1.2.42).

Definition 2.1.1. For a given $k \geq 0$, a multimap $F : X \to C(X)$ is said to be *k-Lipschitz* if for every $x, y \in X$ the following inequality holds:

$$h(F(x), F(y)) \leq k\varrho(x, y).$$

If $k < 1$ then k-Lipschitz multimap F is called *contractive* (*k-contractive*).

The following fixed point theorem holds true (see [315], [110]).

Theorem 2.1.2. *Let (X, ϱ) be a complete metric space, $F : X \to C(X)$ a k-contractive multimap. Let x_0 be a point in X with $\varrho(x_0, F(x_0)) < \delta$. Then F has a fixed point x_* such that $\varrho(x_0, x_*) < \frac{\delta}{1-k}$.*

Proof. Construct a sequence of points $\{x_n\}_{n=0}^{\infty} \subset X$ such that

$$x_n \in F(x_{n-1}) \quad for \quad n = 1, 2, ...,$$

$$\varrho(x_{n+1}, x_n) < k^n \delta \quad for \quad n = 0, 1, ...$$

by the following inductive manner. Let x_0 be a given point, then choose a point x_1 arbitrarily in $F(x_0)$ so that $\varrho(x_0, x_1) < \delta$. Suppose that the points $x_0, x_1, ..., x_n$ of our sequence are constructed already. Then

$$\varrho(x_n, F(x_n)) \leq h(F(x_{n-1}), F(x_n)) \leq k\varrho(x_{n-1}, x_n) < k^n \delta.$$

Therefore, there exists a point $x_{n+1} \in F(x_n)$ such that

$$\varrho(x_n, x_{n+1}) < k^n \delta.$$

It is easy to see that the constructed sequence $\{x_n\}$ is fundamental. In fact,

$$\varrho(x_n, x_{n+p}) \leq \sum_{i=n}^{n+p-1} \varrho(x_i, x_{i+1}) < \sum_{i=n}^{n+p-1} k^i \delta < \frac{k^n \delta}{1-k}.$$

Since the space X is complete, the sequence $\{x_n\}$ converges to a point $x_* \in X$.

Let us show that the point x_* is a fixed point of the multimap F. Indeed,

$$\varrho\left(x_{n+1}, F\left(x_*\right)\right) \leq h\left(F\left(x_n\right), F\left(x_*\right)\right) \leq k\varrho\left(x_n, x_*\right).$$

Hence $\lim_{n \to \infty} \varrho\left(x_{n+1}, F\left(x_*\right)\right) = 0$. Since the set $F\left(x_*\right)$ is closed, we have $x_* \in F\left(x_*\right)$.

As

$$\varrho\left(x_0, x_*\right) \leq \sum_{n=0}^{\infty} \varrho\left(x_n, x_{n+1}\right) < \sum_{n=0}^{\infty} k^n \delta = \frac{\delta}{1-k},$$

we get the desired estimate. ∎

It is obvious that, in contrast to the single-valued case, a contractive multimap can have many fixed points. Moreover, the following assertion holds true.

Theorem 2.1.3. *Let X be a complete metric space, $F : X \to C\left(X\right)$ a k-contractive multimap with $0 < k < \frac{1}{2}$ and $\bar{x}$ a fixed point of F. If $F\left(\bar{x}\right) \neq \{\bar{x}\}$ then F has at least one more fixed point.*

Proof. Let $x_0 \neq \bar{x}$ be an arbitrary point from $F\left(\bar{x}\right)$, then

$$\varrho\left(x_0, F\left(x_0\right)\right) \leq h\left(F\left(\bar{x}\right), F\left(x_0\right)\right) \leq k\varrho\left(\bar{x}, x_0\right) < \delta,$$

where $\delta = k_1\varrho\left(\bar{x}, x_0\right)$, $k < k_1 < \frac{1}{2}$. Then, by Theorem 2.1.2, there exists a fixed point x_* of F such that

$$\varrho\left(x_0, x_*\right) < \frac{\delta}{1-k} = \frac{k_1\varrho\left(\bar{x}, x_0\right)}{1-k} < \varrho\left(\bar{x}, x_0\right).$$

Therefore, $x_* \neq \bar{x}$. ∎

2.1.2 Contractive multimaps depending on a parameter

Let E be a Banach space; X a closed convex subset of E; Y a metric space.

Consider a multimap $F : X \times Y \to Cv\left(X\right)$ satisfying the following conditions:

(i) there exists a number $0 < k < 1$ such that for every $x', x'' \in X$ and $y \in Y$ we have:

$$h\left(F\left(x', y\right), F\left(x'', y\right)\right) \leq k\|x' - x''\|;$$

(ii) the multimap F is lower semicontinuous.

By Theorem 2.1.2 for every $y \in Y$ the multimap $F_y = F(\cdot, y) : X \to Cv(X)$ has a fixed point. Denote $\mathcal{F}(y) = Fix\, F_y = \{x \mid x \in F_y(x)\}$. Consider the so defined multimap $\mathcal{F} : Y \to C(X)$.

Theorem 2.1.4. *Under conditions (i) and (ii), let A be a closed subset of Y, and $f : A \to X$ a continuous map such that $f(y) \in F(f(y), y)$ for every $y \in A$. Then there exists a continuous map $g : Y \to X$ satisfying the following conditions:*

    (a) *g is a continuous selection of the multimap $\mathcal{F}$, i.e., $g(y) \in F(g(y), y)$ for every $y \in Y$;*
    (b) *the map g is a continuous extension of the map f, i.e., $g|_A = f$.*

Proof. Take an arbitrary number $k_0 \in (k, 1)$. Construct a sequence of continuous maps $g_n : Y \to X$, $n = 0, 1, 2, \ldots$ satisfying the following conditions:

    (1) $g_n(y) \in F(g_{n-1}(y), y)$ for every $y \in Y$ and $n = 1, 2, \ldots$;
    (2) there exists a continuous function $r : Y \to \mathbb{R}_+$ such that for each $y \in Y$ and $n = 0, 1, 2, \ldots$ the following inequality holds:

$$\|g_{n+1}(y) - g_n(y)\| \le k_0^n r(y);$$

    (3) $g_n|_A = f$ for every $n = 0, 1, 2, \ldots$.

This sequence will be formed by induction. As the map $g_0 : Y \to X$ we take an arbitrary extension of the map f to the whole space Y, existing by virtue of the Tietze–Dugundji theorem (see Chapter 0). Define the multimap $\Psi_1 : Y \to Cv(E)$ by the formula $\Psi_1(y) = F(g_0(y), y)$. According to Theorem 1.3.11 this multimap is l.s.c. Hence, by Theorem 1.4.2 the multimap Ψ_1 admits a continuous selection $q : Y \to E$. Define a continuous function $r : Y \to \mathbb{R}_+$ as $r(y) = \|g_0(y) - q(y)\| + 1$.

Consider the multimap $Q : Y \to Cv(E)$,

$$Q(y) = \{x \in E \mid \|g_0(y) - x\| \le r(y)\}.$$

It is easy to see that the graph of the multimap $int\, Q$,

$$int\, Q(y) = \{x \in E \mid \|g_0(y) - x\| < r(y)\}$$

is open and hence (see Section 1.3.1) the multimap Q is quasi-open. Notice also that $\Psi_1(y) \cap Q(y) \neq \emptyset$ and, moreover,

$$\Psi_1(y) \cap Q(y) = \overline{\Psi_1(y) \cap int\, Q(y)} \quad \text{for all } y \in Y$$

that yields, by Theorem 1.3.9, that the multimap $\Psi_1 \cap Q \colon Y \to Cv(E)$ is l.s.c. We have also that $f(y) \in (\Psi_1(y) \cap Q(y))$ for every $y \in A$.

Then, applying Corollary 1.4.5, we get a continuous map $g_1 \colon Y \to E$ which is a selection of $\Psi_1 \cap Q$ that coincides with f on A. In other words, g_1 satisfies

$$\|g_1(y) - g_0(y)\| \leq r(y)$$

and hence g_1 satisfies conditions (1)–(3) (for $n = 1$).

Suppose that the maps $g_0, g_1, ..., g_n$ satisfying conditions (1)–(3) are constructed already. Set $\Psi_{i+1}(y) = F(g_i(y), y)$. Then for all $y \in Y$ we have

$$\varrho(g_n(y), \Psi_{n+1}(y)) \leq h(F(g_{n-1}(y), y) ; F(g_n(y), y))$$

$$\leq k\|g_n(y) - g_{n-1}(y)\| \leq k \, k_0^{n-1} r(y) < k_0^n r(y).$$

Consider the multimap $Q_n : Y \to Cv(E)$,

$$Q_n(y) = \{x \in E \mid \|g_n(y) - x\| \leq k_0^n r(y)\}.$$

Since graph of the multimap $int\, Q$,

$$int\, Q_n(y) = \{x \in E \mid \|g_n(y) - x\| < k_0^n \, r(y)\}$$

is open we again come to the conclusion that the multimap Q_n is quasi-open. Since $\Psi_{n+1}(y) \cap Q_n(y) \neq \emptyset$ and

$$\Psi_{n+1}(y) \cap Q_n(y) = \overline{\Psi_{n+1}(y) \cap int\, Q_n(y)} \text{ for all } y \in Y$$

we get that the multimap $\Psi_{n+1} \cap Q_n \colon Y \to Cv(E)$ is l.s.c. We have also that $f(y) \in \Psi_{n+1}(y) \cap Q_n(y)$ for every $y \in A$. Then the multimap Ψ_{n+1} admits a continuous selection g_{n+1} which satisfies conditions (1)–(3), that concludes the construction of the sequence $\{g_n\}$.

Now, let us show that for every $y \in Y$ the sequence $x_n = g_n(y)$ is fundamental. In fact,

$$\|x_{n+p} - x_n\| \leq \|x_{n+1} - x_n\| + \|x_{n+2} - x_{n+1}\| + ... + \|x_{n+p} - x_{n+p-1}\|$$

$$< r(y)\left(k_0^n + k_0^{n+1} + ... + k_0^{n+p-1}\right) < \frac{r(y)}{1 - k_0} k_0^n.$$

Denote $g(y) = \lim\limits_{n \to \infty} g_n(y)$ and prove the continuity of this map. Indeed, let y_0 be an arbitrary point from Y, then in a certain neighborhood V of this point for each $y \in V$ the following relation holds: $\|r(y)\| \leq \|r(y_0)\| + 1$. Then the sequence $\{g_n\}$ uniformly converges to a map g on the set V that provides the continuity of g on this set. Now the continuity of g follows from the arbitrariness of y_0.

Passing to the limit in condition (1) for a fixed y we get the inclusion

$$g(y) \in F(g(y), y).$$

Condition $g|_A = f$ follows from property (3) of the constructed sequence. ∎

As an application of this result, let us consider the Ricceri theorem [369] on the topological structure of a fixed point set for a contractive multimap.

Recall the following notion (see Chapter 0).

Let X be a metric space, A a subspace of X.

Definition 2.1.5. A subspace A is *a retract* of X if there exists a continuous map $r : X \to A$ such that $r(x) = x$ for each $x \in A$. In this case the map r is called *a retraction*.

Retracts "inherit" many topological properties of comprehending spaces. For example, if a space is connected then its retract is also connected (see, for example, [77]).

Theorem 2.1.6. *Let X be a closed convex subset of a Banach space E, $F : X \to Cv(X)$ a contractive multimap. Then the set $\mathcal{F} = Fix\, F$ is the retract of X.*

Proof. Consider the multimap $\widehat{F} : X \times X \to Cv(X)$ defined as $\widehat{F}(x, y) = F(x)$. Take the inclusion map $f : \mathcal{F} \to X$, $f(y) = y$. It is clear that $f(y) \in \widehat{F}(f(y), y)$ for all $y \in \mathcal{F}$. Then, by virtue of Theorem 2.1.4, there exists a continuous map $g : X \to X$ satisfying:

(a) $g(y) \in \widehat{F}(g(y), y) = F(g(y))$, i.e., $g(y) \in \mathcal{F}$ for all $y \in X$;

(b) the map g is a continuous extension of the map f, i.e., for every $y \in \mathcal{F}$ we have $g(y) = f(y) = y$.

This means that the map g is the retraction of the space X onto $\mathcal{F}$. ∎

Corollary 2.1.7. *Under conditions of Theorem 2.1.6 the fixed point set $Fix F$ is connected.*

2.1.3   Equations with surjective linear operators

In this section we will apply theorems on fixed points for contractive multimaps to the study of some classes of equations with linear operators.

Let E_1, E_2 be Banach spaces, $a : D(a) \subset E_1 \to E_2$ a linear operator. Recall the following notion (see, for example, [124], [395], [408]).

Definition 2.1.8. A linear operator a is called *closed*, if its graph is a closed linear subspace of $E_1 \times E_2$, i.e., relations $\{x_n\} \subset D(a)$, $x_n \to x$, $a(x_n) \to y$ yield $x \in D(a)$ and $a(x) = y$.

Let a be a closed linear operator, $Ker(a)$ its kernel. Consider its quotient space $E = E_1/Ker(a)$. It is known that the norm in the space E is defined in the following way: if $[x] = x + Ker(a) \in E$ then $\| [x] \| = \inf\limits_{u \in Ker(a)} \|x + u\|$. Let p be a projection of the space E_1 onto E, $p(x) = [x]$.

Consider the linear operator $a_1 : D(a_1) \subset E \to E_2$, where $D(a_1) = p(D(a))$ and $a_1([x]) = a(x)$. It is clear that the operator a_1 is closed, surjective and has null kernel. Whence the operator a_1 is invertible and the following commutative diagram holds:

$$
\begin{array}{ccc}
D(a) \subset E_1 & \xrightarrow{\ a\ } & E_2 \\
\ \ \ \ p \searrow & & \nearrow a_1 \\
& D(a_1) \subset E &
\end{array}
$$

By the Banach inverse mapping theorem (see, for example, [124], [395], [408]) the operator a_1^{-1} is bounded. Then we have:

$$
\|a_1^{-1}\| = \sup_{y \in E_2} \frac{\|a_1^{-1}(y)\|}{\|y\|} = \sup_{y \in E_2} \left(\frac{\inf\{\|x\| \mid x \in E_1, a(x) = y\}}{\|y\|} \right).
$$

Denote $\|a_1^{-1}\| = \beta(a)$.

Definition 2.1.9. The number $\beta(a)$ is called *the norm* of the multioperator a^{-1} and it will be denoted by $\|a^{-1}\|$.

Consider now the properties of the multioperator $a^{-1} : E_2 \to Cv(E_1)$.

Lemma 2.1.10. *The multioperator* a^{-1} *is Lipschitz with the constant* $\|a^{-1}\|$, *i.e.*,

$$
h\left(a^{-1}(x_1), a^{-1}(x_2)\right) \le \|a^{-1}\| \, \|x_1 - x_2\|.
$$

Proof. For arbitrary $x_1, x_2 \in E_2$ we have

$$
h\left(a^{-1}(x_1), a^{-1}(x_2)\right) = \inf\{ \|z_1 - z_2\| \mid z_1 \in a^{-1}(x_1), z_2 \in a^{-1}(x_2) \}
$$

$$
= \inf\{ \|z_1 - z_2\| \mid z_1 - z_2 \in a^{-1}(x_1 - x_2) \} \le \|a^{-1}\| \, \|x_1 - x_2\|.
$$

■

Now, let $a : D(a) \to E_2$ be a closed linear surjective operator, $f : E_1 \to E_2$ a Lipschitz map, i.e., there exists a constant $c > 0$ such that for every $x_1, x_2 \in E_1$ we have $\|f(x_1) - f(x_2)\| \le c\|x_1 - x_2\|$.

Consider the following equation:

$$a(x) = f(x).$$

Denote by $N(a, f)$ the solution set of this equation, i.e.,

$$N(a, f) = \{x \in E_1 \mid a(x) = f(x)\}.$$

Theorem 2.1.11. *If $c < \frac{1}{\|a^{-1}\|}$ then the set $N(a, f)$ is nonempty and, moreover, it is the retract of the space E_1.*

Proof. It is obvious that the equation under consideration is equivalent to the inclusion $x \in F(x)$, where $F(x) = a^{-1}(f(x))$. Let us show that the multimap F has a fixed point. To this end, notice that it is contractive. In fact,

$$h(F(x), F(y)) = \inf\{\|z_1 - z_2\| \mid z_1 \in F(x), z_2 \in F(y)\}$$
$$= \inf\{\|z_1 - z_2\| \mid a(z_1 - z_2) = f(x) - f(y)\}$$
$$\le \|a^{-1}\| \|f(x) - f(y)\| \le \|a^{-1}\| c\|x - y\|$$

and it remains to use that, by the condition of the theorem, $\|a^{-1}\|c < 1$. Now the validity of the assertion follows from Theorems 2.1.2 and 2.1.6. ∎

It is well known that the property of invertibility of linear operators is stable in the space of all bounded linear operators with respect to small perturbations. The classical proof of this fact is based on the Banach contraction map principle. The stability of the surjectivity property for linear operators with respect to small perturbations also turns out to be the corollary of a fixed point theorem, but for multivalued contractions.

Theorem 2.1.12. *Let $a : D(a) \subset E_1 \to E_2$ be a closed surjective linear operator; $b : E_1 \to E_2$ a bounded linear operator and $\|b\| < \frac{1}{\|a^{-1}\|}$. Then the linear operator $a + b : D(a) \subset E_1 \to E_2$ is surjective.*

Proof. Let y_0 be an arbitrary point of the space E_2. Consider the equation

$$a(x) = y_0 - b(x).$$

It is clear that the map $f(x) = y_0 - b(x)$ is Lipschitz with the constant $c = \|b\| < \frac{1}{\|a^{-1}\|}$. Whence, by Theorem 2.1.11, this equation has a solution, that concludes the proof. ∎

Corollary 2.1.13. *The set of linear bounded surjective operators $S(E_1, E_2)$ is open in the space of bounded linear operators $L(E_1, E_2)$.*

2.1.4 Inequalities of Caristi type and α-contractive multimaps

Let X be a complete metric space, $F : X \to C(X)$ a closed multimap, $\alpha : X \times X \to \mathbb{R}$ a bounded below function and $\gamma_0 = \inf\limits_{(x,y) \in X \times X} \alpha(x,y)$. If A, B are subsets of X, denote

$$\alpha_*(A, B) = \sup_{a \in A} \inf_{b \in B} \alpha(a, b),$$

and

$$h_\alpha(A, B) = \max\{\alpha_*(A, B), \alpha_*(B, A)\}.$$

Consider an assertion which is the development of a multivalued version of the Caristi fixed point theorem [24].

Theorem 2.1.14. *If there exists a number $c > 0$ such that for every points $x \in X$ and $y \in F(x)$ and each $\eta > 0$ there is a point $z \in F(y)$ such that the following inequality holds true:*

$$\alpha(y, z) + c\rho(y, z) \le \alpha(x, y) + \eta \tag{2.1}$$

then F has a fixed point. Moreover, for each points $x \in X$ and $y \in F(x)$ and every $\delta > 0$ there exists such fixed point x_* that

$$\varrho(y, x_*) < \frac{\alpha(x, y) - \gamma_0}{c} + \delta. \tag{2.2}$$

Proof. For an arbitrary $\varepsilon > 0$ construct a sequence $\{x_n\}_{n=0}^{\infty} \in X$ such that:

(i) $x_{n+1} \in F(x_n)$;
(ii) $\alpha(x_n, x_{n+1}) + c\varrho(x_n, x_{n+1}) < \alpha(x_{n-1}, x_n) + \frac{\varepsilon}{2^n}$, $\quad n \ge 1$.

This sequence will be constructed by induction. Take points $x_0 \in X$ and $x_1 = y \in F(x_0)$. Then, by virtue of the conditions of the theorem, there exists a point $x_2 \in F(x_1)$ such that

$$\alpha(x_1, x_2) + c\rho(x_1, x_2) < \alpha(x_0, x_1) + \frac{\varepsilon}{2}.$$

Assume that we constructed points x_0, x_1, x_2, ..., x_n satisfying conditions (i) and (ii). Let us find a point x_{n+1}. Since $x_n \in F(x_{n-1})$ there exists a point $x_{n+1} \in F(x_n)$ such that

$$\alpha(x_n, x_{n+1}) + c\rho(x_n, x_{n+1}) < \alpha(x_{n-1}, x_n) + \frac{\varepsilon}{2^n}.$$

Now, let us show that the sequence $\{\beta_n\}$, where $\beta_n = \alpha(x_{n-1}, x_n)$ is convergent. To this aim, consider the sequence $\left\{\beta_n + \sum\limits_{j=n}^{\infty} \frac{\varepsilon}{2^j}\right\}$. This sequence is decreasing:

$$\left(\beta_n + \sum_{j=n}^{\infty} \frac{\varepsilon}{2^j}\right) - \left(\beta_{n+1} + \sum_{j=n+1}^{\infty} \frac{\varepsilon}{2^j}\right) \geq c\rho(x_n, x_{n+1}) \geq 0.$$

Since the sequence $\left\{\beta_n + \sum\limits_{j=n}^{\infty} \frac{\varepsilon}{2^j}\right\}$ is below bounded, it is convergent that yields the convergence of $\{\beta_n\}$.

Now, let us show that the sequence $\{x_n\}$ is also convergent. From inequality (2.3) we have the estimate

$$\varrho(x_n, x_{n+1}) < \frac{1}{c}(\beta_n + \frac{\varepsilon}{2^n} - \beta_{n+1})$$

for $n \geq 1$. Then

$$\varrho(x_n, x_{n+p}) \leq \sum_{j=n+1}^{n+p} \varrho(x_{j-1}, x_j) < \frac{1}{c} \sum_{j=n+1}^{n+p} \left(\beta_j + \frac{\varepsilon}{2^j} - \beta_{j+1}\right).$$

Further, we get

$$\varrho(x_n, x_{n+p}) < \frac{1}{c}\left((\beta_{n+1} - \beta_{n+p+1}) + \sum_{j=n+1}^{n+p} \frac{\varepsilon}{2^j}\right).$$

The last inequality yields that the sequence $\{x_n\}$ is fundamental and therefore it converges to a point x_*. Since $x_{n+1} \in F(x_n)$, the closedness of F implies that x_* is a fixed point of F.

Let us prove inequality (2.2). Notice that for sufficiently large n we have $\varrho(x_1, x_*) < \sum\limits_{j=1}^{n} \varrho(x_j, x_{j+1}) + \varepsilon$. Then

$$\varrho(x_1, x_*) < \frac{1}{c}\left((\alpha(x_0, x_1) - \alpha(x_n, x_{n+1})) + \sum_{j=1}^{n} \frac{\varepsilon}{2^j}\right) + \varepsilon$$

$$< \frac{1}{c}\left(\alpha(x_0, x_1) - \gamma_0 + \varepsilon\right) + \varepsilon = \frac{\alpha(x_0, x_1) - \gamma_0}{c} + \frac{\varepsilon(c+1)}{c}.$$

Since a number ε is arbitrary, we can assume $\frac{\varepsilon(c+1)}{c} < \delta$ that implies the validity of (2.2) and concludes the proof. ∎

Definition 2.1.15. A multimap F is said to be *α-contractive* if there exists a number $k \in (0,1)$ such that for every $x, y \in X$ the following inequality holds:

$$h_\alpha(F(x), F(y)) \le k\,\alpha(x, y), \tag{2.4}$$

Let us apply Theorem 2.1.14 to the study of fixed points of α-contractive multimaps.

Theorem 2.1.16. *Let $F : X \to C(X)$ be a closed multimap satisfying the following conditions:*

(1) *there exists a number $q > 0$ such that for every $x, y \in X$ we have the inequality $\varrho(x, y) \le q\,\alpha(x, y)$;*
(2) *F is α-contractive with a constant k.*

Then the multimap F has a fixed point and, moreover, for every number $\delta > 0$ there exists a fixed point x_* such that for each $x \in X$ and $y \in F(x)$ we have

$$\varrho(y, x_*) < \frac{qk\,(\alpha(x, y) - \gamma_0)}{(1 - k)} + \delta. \tag{2.5}$$

Proof. Take arbitrary $x \in X$, $y \in F(x)$ then by inequality (2.4) and the definition of h_α, for every positive number η there exists a point $z \in F(y)$ such that $\alpha(y, z) \le k\,\alpha(x, y) + \eta$. Since $k \in (0,1)$ we can represent it as $k = \frac{1}{1+c}$, where $c > 0$. Then

$$\alpha(y, z) + c\,\alpha(y, z) = \alpha(y, z) + \frac{1 - k}{k}\,\alpha(y, z) \le \alpha(x, y) + \eta.$$

Whence, we get the inequality

$$\alpha(y, z) + \frac{(1 - k)}{qk}\varrho(y, z) \le \alpha(x, y) + \eta.$$

Now the validity of the assertion follows from Theorem 2.1.14. ∎

The above theorem naturally generalizes the Nadler theorem (Theorem 2.1.2). Notice also that from Theorem 2.1.4 the next "non-symmetric" version of Theorem 2.1.16 follows.

Corollary 2.1.17. *Let $F : X \to C(X)$ be a closed multimap satisfying the following conditions:*

(1) *there exists a number $q > 0$ such that for every $x, y \in X$ we have $\varrho(x, y) \le q\,\alpha(x, y)$;*

(2) *there exists a number $k \in (0,1)$ such that for every $x \in X$, $y \in F(x)$ the next inequality holds:*

$$\alpha_*(y, F(y)) \leq k\,\alpha(x, y).$$

Then F has a fixed point and, moreover, for each points $x \in X$, $y \in F(x)$ and every $\delta > 0$ there exists a fixed point x_* such that

$$\varrho(y, x_*) < \frac{qk\,(\alpha(x,y)) - \gamma_0)}{(1 - k)} + \delta.$$

Let us present an example of a multimap satisfying the conditions of Corollary 2.1.17 which is not α-contractive.

Example 2.1.18. Take $X = [-2, 2] \times [-2, 2] \subset \mathbb{R}^2$ and choose points $x = (x_1, x_2), y = (y_1, y_2) \in X$. The metric in X is given by $\varrho(x, y) = \max\{|x_1 - y_1|; |x_2 - y_2|\}$. Define the multimap $F : X \to C(X)$ in the following way: if $x = (x_1, x_2)$ then

$$F(x) = \begin{cases} (\frac{1}{2}x_1, 1) & \text{if } x_1 > 0, \\ (\frac{1}{2}x_1, -1), & \text{if } x_1 < 0, \\ \{(0, 1); (0, -1)\}, & \text{if } x_1 = 0. \end{cases}$$

It is easy to verify **(do it!)** that this multimap is not contractive but it satisfies the conditions of Corollary 2.1.17 if we will take the metric ϱ as the function α.

2.1.5   *Fixed points of weakly α-contractive multimaps*

In 1936 V.V. Nemytskii [317] noted that the conclusion of the contraction map principle remains true for a map f transforming a metric space (X, ϱ) into its compact part and satisfying a weak contraction condition: $\varrho(f(x), f(y)) < \varrho(x, y)$, $\forall x, y \in X, x \neq y$. Consider the generalization of this theorem to the case of weakly α-contractive multimaps.

Let X be a compact metric space, $\alpha : X \times X \to \mathbb{R}$ a lower semicontinuous function, $F : X \to K(X)$ a closed multimap.

Theorem 2.1.19. *If for every $x \in X$ and $y \in F(x)$, $x \neq y$ the following inequality holds:*

$$\alpha_*(y, F(y)) < \alpha(x, y)$$

then the multimap F has a fixed point.

Proof. Consider the function $\beta : X \to \mathbb{R}$ given as

$$\beta(x) = \min_{u \in F(x)} \alpha(x, u).$$

Let us show that the function β is also lower semicontinuous. In fact, let x_0 be an arbitrary point from X and $\{x_n\} \to x_0$. Let $\beta(x_n) = \alpha(x_n, u_n)$. Without loss of generality we can assume that $\{u_n\} \to u_\star$. Then, from the closedness of F we get $u_\star \in F(x_0)$. Hence $\beta(x_n) \to \alpha(x_0, u_\star)$ and $\beta(x_0) \leq \alpha(x_0, u_\star)$ yielding the lower semicontinuity of the function β.

Then the function β achieves its minimum on the set X :

$$\beta(x_\star) = \min_{x \in X} \beta(x).$$

Let us show that $x_\star \in F(x_\star)$. Assume the contrary: $x_\star \notin F(x_\star)$. Then $\beta(x_\star) = \alpha(x_\star, y_\star)$ and $y_\star \neq x_\star$ and

$$\alpha_\star(y_\star, F(y_\star)) < \alpha(x_\star, y_\star) = \beta(x_\star).$$

This means that there exists a point $z_\star \in F(y_\star)$ such that $\alpha(y_\star, z_\star) < \alpha(x_\star, y_\star)$. Then we have $\beta(y_\star) < \beta(x_\star)$. The obtained contradiction concludes the proof of the theorem. ∎

2.2 Topological degree of compact multivalued vector fields

> Be yourself. A horse without the lancer is still a horse, a lancer without the horse is just a man.
>
> *Stanislaw Jerzy Lec*

The notion of the topological degree (rotation) is a convenient and effective tool which is applied to fixed point problems for multimaps as well as to other questions. In this section we define the notion of a relative topological degree for a compact multivalued vector field with convex values in a Banach space, describe its main properties and present applications to fixed point theorems.

Everywhere in the sequel E is a real Banach space.

Let $X \subset E$. Each multimap $F : X \to P(E)$ defines a multimap $\Phi : X \to P(E)$,

$$\Phi(x) = x - F(x)$$

called *the multivalued vector field* or *the multifield* corresponding to the multimap F. Denoting $i : X \to E$ the inclusion map we will write

$$\Phi = i - F.$$

If Λ is the space of parameters and $G : X \times \Lambda \to P(E)$ is a family of multimaps then $\Psi : X \times \Lambda \to P(E)$ given as

$$\Psi(x, \lambda) = x - G(x, \lambda),$$

is called *the family of multifields*.

A point $x \in X$ such that

$$0 \in \Phi(x)$$

is called *a zero point* of the multifield Φ. It is clear that zero points of a multifield $\Phi = i - F$ are just fixed points of the multimap F. If a multifield Φ has no zero points we will call it *nondegenerate*.

Let us agree also to the following terminology. For convenience, in what follows a upper semicontinuous and compact multimap F will be merely called *compact*. Analogously, a single-valued map which is continuous and compact will be called compact. The same term "compact" will be used for the corresponding vector fields (multivalued or single-valued).

Notice that if $F_0, F_1 : X \to K(E)$ are compact multimaps then (see Section 1.3.2) the multimap $G : X \times [0, 1] \to K(E)$,

$$G(x, \lambda) = \lambda F_1(x) + (1 - \lambda) F_0(x)$$

is also compact.

Now, let $\mathbf{K} \subset E$ be a nonempty convex closed set and $U_{\mathbf{K}}$ is a nonempty open (in the relative topology) subset of $\mathbf{K}$. The closure and the boundary of the set $U_{\mathbf{K}}$ in the same relative topology of the space $\mathbf{K}$ will be denoted by $\overline{U}_{\mathbf{K}}$ and $\partial U_{\mathbf{K}}$ respectively. At first, we will suppose that $\partial U_{\mathbf{K}}$ is nonempty.

It is known that to each compact single-valued map $f : \partial U_{\mathbf{K}} \to \mathbf{K}$ which is fixed points free, one can associate an integer characteristic, *the relative topological degree*

$$deg_{\mathbf{K}}(\varphi, \partial U_{\mathbf{K}})$$

of the corresponding vector field $\varphi = i - f$, $\varphi(x) = x - f(x)$ (see [64], [65], [198]). This characteristic has the following principal properties.

Property 2.2.1 (The normalization property). *If $f(x) \equiv x_0$ for all $x \in \partial U_{\mathbf{K}}$ then*

$$deg_{\mathbf{K}}(\varphi, \partial U_{\mathbf{K}}) = \begin{cases} 1, & \text{if } x_0 \in U_{\mathbf{K}}, \\ 0, & \text{if } x_0 \notin U_{\mathbf{K}}. \end{cases}$$

Property 2.2.2 (The homotopy invariance). *Assume that compact fields $\varphi_0 = i - f_0$ and $\varphi_1 = i - f_1$ are homotopic ($\varphi_0 \sim \varphi_1$, i.e., there exists a compact map $g : \partial U_{\mathbf{K}} \times [0, 1] \to \mathbf{K}$ such that*

(1) $x \neq g(x, \lambda)$ *for all* $x \in \partial U_{\mathbf{K}}$, $\lambda \in [0, 1]$;

(2) $g(\cdot, 0) = f_0$, $g(\cdot, 1) = f_1$.

Then $deg_{\mathbf{K}}(\varphi_0, \partial U_{\mathbf{K}}) = deg_{\mathbf{K}}(\varphi_1, \partial U_{\mathbf{K}})$.

Property 2.2.3 (The additive dependence on a domain). *Let* $\{U_{j\mathbf{K}}\}_{j \in J}$ *be a finite family of relatively open disjoint subsets of* $U_{\mathbf{K}}$; *a map* $f : \bar{U}_{\mathbf{K}} \to \mathbf{K}$ *compact and fixed point free on the set* $\overline{U}_{\mathbf{K}} \setminus \bigcup_{j \in J} U_{j\mathbf{K}}$. *Then*

$$deg_{\mathbf{K}}(i - f, \partial U_{\mathbf{K}}) = \sum_{j \in J} deg_{\mathbf{K}}(i - f, \partial U_{j\mathbf{K}}).$$

Property 2.2.4 (The map restriction principle). *Let* $\mathbf{K}_1$ *be a nonempty convex closed subset of* E, $\mathbf{K}_1 \subset \mathbf{K}$ *and* $\partial U_{\mathbf{K}_1} = \partial(U \cap \mathbf{K}_1) \neq \emptyset$. *If a compact map* $f : \partial U_{\mathbf{K}} \to \mathbf{K}$ *is fixed point free and* $f(\partial U_{\mathbf{K}}) \subset \mathbf{K}_1$ *then*

$$deg_{\mathbf{K}}(i - f, \partial U_{\mathbf{K}}) = deg_{\mathbf{K}_1}(i - f, \partial U_{\mathbf{K}_1}).$$

Property 2.2.5 (The odd field property). *Let* $\mathbf{K}$ *and* U *be symmetric with respect to the origin and* $0 \in U_{\mathbf{K}}$. *Assume that* $f : \partial U_{\mathbf{K}} \to \mathbf{K}$ *is a compact, fixed point free map which is odd, i.e.,*

$$f(-x) = -f(x)$$

for all $x \in \partial U_{\mathbf{K}}$. *Then the topological degree* $deg_{\mathbf{K}}(i - f, \partial U_{\mathbf{K}})$ *is an odd number.*

Property 2.2.6 (The fixed point property). *If a compact map* $f : \overline{U}_{\mathbf{K}} \to \mathbf{K}$ *is fixed point free on* $\partial U_{\mathbf{K}}$ *and*

$$deg_{\mathbf{K}}(i - f, \partial U_{\mathbf{K}}) \neq 0$$

then f *has a fixed point* x *in* $U_{\mathbf{K}}$, *i.e., there exists* $x \in U_{\mathbf{K}}$ *such that* $x = f(x)$.

Remark 2.2.7. In the "absolute" case $\mathbf{K} = \mathbf{E}$ we get the classical Leray – Schauder topological degree $deg(i - f, \partial U)$ of a compact vector field, which construction is described in numbers of works (see, e.g., [21], [115], [262], [414]). The transfer from the classical degree to the relative one may be realized by many ways. One of the simplest consists of taking a retraction $r : E \to \mathbf{K}$ (which exists by the Tietze–Dugundji theorem, see Chapter 0) and then setting

$$deg_{\mathbf{K}}(i - f, \partial U_{\mathbf{K}}) := deg\left(i - \tilde{f}, \partial \tilde{U}\right),$$

where $\widetilde{U} = r^{-1}(U_{\mathbf{K}})$, $\widetilde{f} = f \circ r$. It can be verified that so defined value of the degree does not depend on the choice of the retraction r.

We start now with the construction of the relative topological degree for a compact multivalued vector field. To this aim, we introduce the following concepts.

Definition 2.2.8. Compact multimaps $F_0, F_1 : \partial U_{\mathbf{K}} \to Kv(\mathbf{K})$ and the corresponding multifields $\Phi_0 = i - F_0$, $\Phi_1 = i - F_1$ are called *homotopic*

$$\Phi_0 \sim \Phi_1$$

if there exists a compact multimap $G : \partial U_{\mathbf{K}} \times [0,1] \to Kv(\mathbf{K})$ such that

(1) *Fix* $G(\cdot, \lambda) = \emptyset$ for all $\lambda \in [0,1]$;
(2) $G(\cdot, 0) = F_0$, $G(\cdot, 1) = F_1$.

It is easy to see (**verify it!**) that the relation of homotopy is the equivalence relation on the class of all nondegenerate compact multifields defined on $\partial U_{\mathbf{K}}$.

Definition 2.2.9. A compact single-valued vector field $\varphi = i - f$, $f : \partial U_{\mathbf{K}} \to \mathbf{K}$ is called *a single-valued homotopic approximation* of a compact multifield $\Phi = i - F$, $F : \partial U_{\mathbf{K}} \to Kv(\mathbf{K})$, if $\varphi \sim \Phi$.

To prove the existence of single-valued homotopic approximations we need the following assertion.

Lemma 2.2.10. *Let $X \subset E$ be a topological subspace; Δ a compact metric space; $F : X \times \Delta \to Kv(E)$ a compact multimap such that $F(\cdot, \mu)$ is fixed point free on a closed subset $X_1 \subset X$ for all $\mu \in \Delta$. Then for all sufficiently small $\varepsilon > 0$, if $f_\varepsilon : X \times \Delta \to E$ is a regular ε-approximation of F then the multimap $G : X \times \Delta \times [0,1] \to Kv(E)$,*

$$G(x, \mu, \lambda) = \lambda f_\varepsilon(x, \mu) + (1 - \lambda) F(x, \mu)$$

is such that $x \notin G(x, \mu, \lambda)$ for all $(x, \mu, \lambda) \in X_1 \times \Delta \times [0,1]$.

Proof. Supposing the contrary, we will have sequences $\{\varepsilon_n\}_{n=1}^\infty$, $\varepsilon_n > 0$, $\varepsilon_n \to 0$; $\{x_n\}_{n=1}^\infty \subset X_1$, $\{\mu_n\}_{n=1}^\infty \subset \Delta$ and $\{\lambda_n\}_{n=1}^\infty \subset [0,1]$ such that

$$x_n \in \lambda_n f_{\varepsilon_n}(x_n, \mu_n) + (1 - \lambda_n) F(x_n, \mu_n)$$

for all $n = 1, 2, ...$, where $f_{\varepsilon_n} : X \times \Delta \to E$ are regular ε_n-approximations of F. The sequences $\{f_{\varepsilon_n}(x_n, \mu_n)\}$ and $\{x_n\}$ are contained in a compact set

$\overline{co}F(X)$ and hence we may assume, without loss of generality, that they are convergent: $f_{\varepsilon_n}(x_n, \mu_n) \to v_0$, $x_n \to x_0 \in X_1$. It is clear that the sequences $\{\mu_n\}$ and $\{\lambda_n\}$ can also be regarded as convergent: $\mu_n \to \mu_0 \in \Delta$, $\lambda_n \to \lambda_0 \in [0,1]$.

For points x_n we have the representation

$$x_n = \lambda_n f_{\varepsilon_n}(x_n, \mu_n) + (1 - \lambda_n) y_n,$$

where $y_n \in F(x_n, \mu_n)$. By virtue of compactness of F, the sequence $\{y_n\}$ can also be assumed convergent: $y_n \to y_0$, and the closedness of F implies $y_0 \in F(x_0, \mu_0)$. Each point of the form $((x_n, \mu_n), f_{\varepsilon_n}(x_n, \mu_n)) \in X_1 \times \Delta \times E$ lies in the ε_n-neighborhood of the graph Γ_F, and so the limit point (x_0, μ_0, v_0) belongs to this graph, i.e., $v_0 \in F(x_0, \mu_0)$. Passing to the limit, we get

$$x_0 = \lambda_0 v_0 + (1 - \lambda_0) y_0 \in F(x_0, \mu_0),$$

that contradicts to the absence of fixed points of $F(\cdot, \mu_0)$ on X_1. ∎

Theorem 2.2.11. *Each nondegenerate compact multifield $\Phi = i - F$, $F : \partial U_{\mathbf{K}} \to Kv(\mathbf{K})$ admits a single-valued homotopic approximation.*

Proof. From the previous lemma it follows that as a single-valued homotopic approximation of Φ we may take a field $\varphi = i - f$, where $f : \partial U_{\mathbf{K}} \to \mathbf{K}$ is an arbitrary regular ε-approximation of the multimap F and $\varepsilon > 0$ is sufficiently small.

Moreover, we can estimate the smallness of an appropriate ε. In fact it is easy to see **(verify it!)** that the set $\Phi(\partial U_{\mathbf{K}})$ is closed and hence its distance δ from the origin is positive. Then, as ε we can take a number satisfying the relation $0 < \varepsilon < \delta/2$. Indeed, let $f_\varepsilon : \partial U_{\mathbf{K}} \to \mathbf{K}$ be a regular ε-approximation of F satisfying also condition (i) of Theorem 1.4.11. The homotopy between $\varphi_\varepsilon = i - f_\varepsilon$ and Φ is generated by the family of multimaps $G : \partial U_{\mathbf{K}} \times [0,1] \to Kv(\mathbf{K})$ of the form

$$G(x, \lambda) = \lambda f_\varepsilon(x) + (1 - \lambda) F(x).$$

The compactness of G is evident. It remains to show only that $x \notin G(x, \lambda)$ for all $x \in \partial U_{\mathbf{K}}$, $\lambda \in [0,1]$.

Supposing the contrary, we will have a point $x_0 \in \partial U_{\mathbf{K}}$ and a number $\lambda_0 \in [0,1]$ such that $x_0 \in \lambda_0 f_\varepsilon(x_0) + (1 - \lambda_0) F(x_0)$. According to condition (i) let us find a point $x' \in \partial U_{\mathbf{K}}$ such that $\|x' - x_0\| < \varepsilon$ and $f_\varepsilon(x_0) \cup F(x_0) \subset F_\varepsilon(x')$.

But then $x_0 \in F_\varepsilon(x')$ and hence the distance from x' to $F(x')$ is less than $2\varepsilon < \delta$ in contradiction to the choice of δ. ∎

We can introduce now the following concept.

Definition 2.2.12. *The relative topological degree*

$$deg_{\mathbf{K}}\left(\Phi, \partial U_{\mathbf{K}}\right)$$

of a nondegenerate compact multifield $\Phi = i - F$,

$$F : \partial U_{\mathbf{K}} \to Kv\left(\mathbf{K}\right)$$

is the relative topological degree $deg_{\mathbf{K}}\left(\varphi, \partial U_{\mathbf{K}}\right)$ of its arbitrary single-valued homotopic approximation φ.

Let us show that this notion is well defined, i.e., the topological degree $deg_{\mathbf{K}}\left(\Phi, \partial U_{\mathbf{K}}\right)$ does not depend on the choice of a single-valued homotopic approximation.

Lemma 2.2.13. *Let $\varphi_0 = i - f_0$ and $\varphi_1 = i - f_1$ be single-valued homotopic approximations of a nondegenerate compact multifield $\Phi = i - F$, $F : \partial U_{\mathbf{K}} \to Kv\left(\mathbf{K}\right)$. Then φ_0 and φ_1 are homotopic in the class of single-valued compact vector fields.*

Proof. From Definition 2.2.9 it follows that the fields φ_0 and φ_1 are homotopic in the class of compact multifields. Let $G : \partial U_{\mathbf{K}} \times [0,1] \to Kv\left(\mathbf{K}\right)$ be a compact fixed point free multimap realizing this homotopy, i.e., $G\left(\cdot, 0\right) = f_0$, $G\left(\cdot, 1\right) = f_1$.

According to Lemma 2.2.10, there exists a regular ε-approximation $g : \partial U_{\mathbf{K}} \times [0,1] \to \mathbf{K}$ of the multimap G such that

$$x \notin \lambda g\left(x, \mu\right) + \left(1 - \lambda\right) G\left(x, \mu\right)$$

for all $x \in \partial U_{\mathbf{K}}$, $\mu \in [0,1]$, $\lambda \in [0,1]$.

Then the required homotopy connecting φ_0 and φ_1 is generated by the compact map $h : \partial U_{\mathbf{K}} \times [0,1] \to \mathbf{K}$,

$$h\left(x, \nu\right) = \begin{cases} 3\nu g\left(x, 0\right) + \left(1 - 3\nu\right) f_0\left(x\right), & 0 \leq \nu \leq \frac{1}{3}; \\ g\left(x, 3\nu - 1\right), & \frac{1}{3} \leq \nu \leq \frac{2}{3}; \\ \left(3 - 3\nu\right) g\left(x, 1\right) + \left(3\nu - 2\right) f_1\left(x\right), & \frac{2}{3} \leq \nu \leq 1, \end{cases}$$

∎

From this statement and the homotopy invariance property 2.2.2 of the degree it follows that the topological degrees of all single-valued homotopic approximations of the multifield Φ are equal, justifying Definition 2.2.12.

Consider the main properties of the defined characteristic.

Theorem 2.2.14 (The normalization property). *If* $F(x) \equiv A$ *for all* $x \in \partial U_{\mathbf{K}}$, *where* $A \subset E$ *is a compact convex set then*

$$deg_{\mathbf{K}}(i - F, \partial U_{\mathbf{K}}) = \begin{cases} 1, & \text{if } A \subset U_{\mathbf{K}}, \\ 0, & \text{if } A \cap \overline{U}_{\mathbf{K}} = \emptyset. \end{cases}$$

Proof. It is sufficient to take as a single-valued homotopic approximation the field $\varphi = i - f$, where $f(x) \equiv x_0 \in A$. ∎

Theorem 2.2.15 (The homotopy invariance). *If* $\Phi_0 \sim \Phi_1$, *then*

$$deg_{\mathbf{K}}(\Phi_0, \partial U_{\mathbf{K}}) = deg_{\mathbf{K}}(\Phi_1, \partial U_{\mathbf{K}}).$$

Proof. This assertion immediately follows from Definition 2.2.12 and the transitivity property of the homotopy relation. ∎

This property serves as the base for a method which is often used while the calculation of the topological degree: a multifield whose degree it is necessary to evaluate is transformed homotopically to a more simple one, for which the degree can be found easily.

In this connection, notice two convenient conditions for the homotopy of multifields. The first one demonstrates the stability of the topological degree with respect to small perturbations of a multifield.

Theorem 2.2.16. *Let* $\mathbf{K}$ *be a convex cone in* E (*i.e.,* $\lambda \mathbf{K} \subset \mathbf{K}$ *for each* $\lambda > 0$ *and* $\mathbf{K} + \mathbf{K} \subset \mathbf{K}$). *Let a compact multifield* $\Phi_0 = i - F_0$, $F_0 : \partial U_{\mathbf{K}} \to Kv(\mathbf{K})$ *be nondegenerate; a compact multimap* $\widetilde{F} : \partial U_{\mathbf{K}} \to Kv(\mathbf{K})$ *be such that that for all* $x \in \partial U_{\mathbf{K}}$ *we have*

$$\|\widetilde{F}(x)\| < \min_{z \in \Phi_0(x)} \|z\|.$$

Then the multifield $\Phi_1 = i - \left(F_0 + \widetilde{F}\right)$ *corresponding to the multimap* $F_0 + \widetilde{F}$ *is nondegenerate and homotopic to* Φ_0, *and hence*

$$deg_{\mathbf{K}}(\Phi_1, \partial U_{\mathbf{K}}) = deg_{\mathbf{K}}(\Phi_0, \partial U_{\mathbf{K}}).$$

Proof. The homotopy of the multifields Φ_0 and Φ_1 is provided by the compact multimap $G : \partial U_{\mathbf{K}} \times [0,1] \to Kv(\mathbf{K})$, $G(x, \lambda) = F_0(x) + \lambda \widetilde{F}(x)$. In fact, if for any $x_0 \in \partial U_{\mathbf{K}}$ and $\lambda_0 \in [0,1]$ we have $x_0 \in G(x_0, \lambda_0)$ then $x_0 = y_0 + \lambda_0 \widetilde{y}$, where $y_0 \in F_0(x_0)$, $\widetilde{y} \in \widetilde{F}(x_0)$. But then $x_0 - y_0 \in \Phi_0(x_0)$ and $\|x_0 - y_0\| = \|\lambda_0 \widetilde{y}\| \leq \|\widetilde{y}\|$ contrary to the condition of the theorem. ∎

Theorem 2.2.17 (An analog of the Poincare–Bohl theorem).
Suppose that nondegenerate compact multifields $\Phi_0 = i - F_0$, $\Phi_1 = i - F_1$ do not admit opposite directions on $\partial U_{\mathbf{K}}$, i.e.,

$$\frac{z_0}{\|z_0\|} \neq -\frac{z_1}{\|z_1\|}$$

for all $x \in \partial U_{\mathbf{K}}$, $z_0 \in \Phi_0(x)$, $z_1 \in \Phi_1(x)$. Then $\Phi_0 \sim \Phi_1$ and hence

$$deg_{\mathbf{K}}(\Phi_0, \partial U_{\mathbf{K}}) = deg_{\mathbf{K}}(\Phi_1, \partial U_{\mathbf{K}}).$$

Proof. The homotopy of the multifields Φ_0 and Φ_1 is provided by the compact multimap $G : \partial U_{\mathbf{K}} \times [0,1] \to Kv(\mathbf{K})$, $G(x, \lambda) = \lambda F_1(x) + (1 - \lambda) F_0(x)$. In fact, if $x_0 \in G(x_0, \lambda_0)$ for any $x_0 \in \partial U_{\mathbf{K}}$ and $\lambda_0 \in (0, 1)$ then $x_0 = \lambda_0 y_1 + (1 - \lambda_0) y_0$, where $y_1 \in F_1(x_0)$, $y_0 \in F_0(x_0)$. But this is equivalent to the equality

$$x_0 - y_0 = -\frac{\lambda_0}{1 - \lambda_0}(x_0 - y_1),$$

contrary to the condition of the theorem. ∎

Consider now other properties of the relative topological degree.

Theorem 2.2.18 (The additive dependence on a domain). *Let $\{U_{j\mathbf{K}}\}_{j \in J}$ be a finite family of relatively open disjoint subsets of $U_{\mathbf{K}}$; a multimap*

$$F : \overline{U}_{\mathbf{K}} \to Kv(\mathbf{K})$$

is compact and fixed point free on the set $\overline{U}_{\mathbf{K}} \setminus \bigcup_{j \in J} U_{j\mathbf{K}}$. Then

$$deg_{\mathbf{K}}(i - F, \partial U_{\mathbf{K}}) = \sum_{j \in J} deg_{\mathbf{K}}(i - F, \partial U_{j\mathbf{K}}).$$

Proof. Let $f : \bar{U}_{\mathbf{K}} \to \mathbf{K}$ be a regular ε-approximation of the multimap F. From Lemma 2.2.10 it follows that if $\varepsilon > 0$ is sufficiently small then f has no fixed points on $\overline{U}_{\mathbf{K}} \setminus \bigcup_{j \in J} U_{j\mathbf{K}}$ and the restriction of f to each of the boundaries $\partial U_{\mathbf{K}}$, $\partial U_{j\mathbf{K}}$, $j \in J$ is the homotopic approximation for the restriction of F to the same boundary. Then the assertion follows from Property 2.2.3 for single-valued fields. ∎

Theorem 2.2.19 (The map restriction principle). *Let*
$\mathbf{K_1}$ *be a nonempty convex closed subset of* E, $\mathbf{K_1} \subset \mathbf{K}$ *and* $\partial U_{\mathbf{K_1}} = \partial (U \cap \mathbf{K_1}) \neq \emptyset$. *If a compact multimap* $F : \partial U_{\mathbf{K}} \to Kv(\mathbf{K})$ *is fixed point free and* $F(\partial U_{\mathbf{K}}) \subset \mathbf{K_1}$ *then*

$$deg_{\mathbf{K}}(i - F, \partial U_{\mathbf{K}}) = deg_{\mathbf{K_1}}(i - F, \partial U_{\mathbf{K_1}}).$$

Proof. According to Lemma 2.2.10, we can choose a regular ε-approximation $f : \partial U_{\mathbf{K}} \to \mathbf{K_1}$ of the multimap F such that it will define the degree $deg_{\mathbf{K}}(i - F, \partial U_{\mathbf{K}})$ and its restriction to $\partial U_{\mathbf{K_1}}$ will be the homotopic approximation of F on this set. Then the assertion is proved by the application of Property 2.2.4 to f. ∎

We can now formulate the basic fixed point principle.

Theorem 2.2.20 (The fixed point property). *Let a compact multimap* $F : \overline{U}_{\mathbf{K}} \to Kv(\mathbf{K})$ *is fixed point free on* $\partial U_{\mathbf{K}}$ *and*

$$deg_{\mathbf{K}}(i - F, \partial U_{\mathbf{K}}) \neq 0.$$

Then F *has a fixed point in* $U_{\mathbf{K}}$.

Proof. In fact, if the multimap F has no fixed points then, applying Lemma 2.2.10 we can construct such regular ε-approximation $f : \overline{U}_{\mathbf{K}} \to \mathbf{K}$ of the multimap F which also has no fixed points and its restriction to the boundary $\partial U_{\mathbf{K}}$ is a single-valued homotopic approximation of the restriction of F to the same boundary. Then, applying Property 2.2.6, we get:

$$deg_{\mathbf{K}}(i - F, \partial U_{\mathbf{K}}) = deg_{\mathbf{K}}(i - f, \partial U_{\mathbf{K}}) = 0,$$

contrary to the assumption. ∎

This principle allows to give a simple and geometrically clear proof of many fixed point results used in applications. Consider some examples.

Theorem 2.2.21. *Let the set* $U_{\mathbf{K}}$ *be convex,* $F : \overline{U}_{\mathbf{K}} \to Kv(\mathbf{K})$ *a compact multimap such that*

$$F(x) \cap \overline{U}_{\mathbf{K}} \neq \emptyset \text{ for all } x \in \partial U_{\mathbf{K}}.$$

Then $FixF \neq \emptyset$.

Proof. If the boundary $\partial U_{\mathbf{K}}$ contains a fixed point of F, the assertion is proved. Let $FixF \cap \partial U_{\mathbf{K}} = \emptyset$. From Corollary 1.3.4 it follows that the multimap $F_1 : \partial U_{\mathbf{K}} \to Kv(\mathbf{K})$, $F_1(x) = F(x) \cap \overline{U}_{\mathbf{K}}$ is compact. Let $x_0 \in U_{\mathbf{K}}$ be an arbitrary point, then it is easy to see that the multimap $G : \partial U_{\mathbf{K}} \times [0,1] \to Kv(\mathbf{K})$,

$$G(x,\lambda) = \begin{cases} (1-2\lambda)F(x) + 2\lambda F_1(x), & 0 \leq \lambda \leq \tfrac{1}{2}; \\ (2-2\lambda)F_1(x) + (2\lambda-1)x_0, & \tfrac{1}{2} \leq \lambda \leq 1, \end{cases}$$

generates the homotopy of the multifield $i - F$ and the field $i - F_0$, where $F_0(x) \equiv x_0$. Applying now the properties of the homotopy invariance and normalization we get

$$deg_{\mathbf{K}}(i - F, \partial U_{\mathbf{K}}) = deg_{\mathbf{K}}(i - F_0, \partial U_{\mathbf{K}}) = 1$$

and it remains to use Theorem 2.2.20.   ∎

The corollary of this result is the well known Bohnenblust–Karlin fixed point theorem [63]:

Theorem 2.2.22. *If M is a convex closed subset of E and a multimap $F : M \to Kv(M)$ is compact then*

$$FixF \neq \emptyset.$$

Proof. Let us take $\mathbf{K} = \mathbf{E}$ and an arbitrary open convex bounded set $U \subset E$ containing the compact set $M_1 = \overline{co}F(M) \subset M$. Consider now the retraction $r : \overline{U} \to M_1$ and the multimap $F_1 : \overline{U} \to Kv(M_1)$ defined as $F_1 = F \circ r$. Such a map is compact (see Theorem 1.3.11) and obviously satisfies conditions of Theorem 2.2.21. Therefore $FixF_1 \neq \emptyset$, but if $x \in FixF_1$ then $x \in M_1$ and hence $r(x) = x$. So we get $x \in F_1(x) = F(r(x)) = F(x)$, i.e., x is a fixed point of F.   ∎

For a finite-dimensional space E this statement is the generalization to the case of multivalued maps of a classical Brouwer fixed point theorem (see Chapter 0). It was proved by S. Kakutani [220]. Its generalization to the case of a locally convex linear topological space E was given by I.L. Glicksberg [182] and Ky Fan [132].

Remark 2.2.23. Now we can include into the general scheme also the case $\partial U_{\mathbf{K}} = \emptyset$. In case $U_{\mathbf{K}} = \emptyset$ we set $deg_{\mathbf{K}}(i - F, \partial U_{\mathbf{K}}) = 0$ and in case $U_{\mathbf{K}} = \mathbf{K}$ it is natural to assume that

$$deg_{\mathbf{K}}(i - F, \partial U_{\mathbf{K}}) = 1.$$

Let us present one more fixed point theorem which can be proved by the evaluation of the topological degree.

Theorem 2.2.24. *Let $\partial U_{\mathbf{K}} \neq \emptyset$ and $F : \overline{U}_{\mathbf{K}} \to Kv(\mathbf{K})$ be a compact multimap, $\Phi = i - F$ be the corresponding multifield. Suppose that for every $x \in \partial U_{\mathbf{K}}$ we have*

$$\Phi(x) \cap L_x^a = \emptyset,$$

where $a \in U_{\mathbf{K}}$ is a certain point and

$$L_x^a = \{z \in E \mid z = \mu(x - a), \ \mu \leq 0\}.$$

Then $deg_{\mathbf{K}}(\Phi, \partial U_{\mathbf{K}}) = 1$ and hence $FixF \neq \emptyset$.

Proof. The restriction of the multifield Φ to $\partial U_{\mathbf{K}}$ and the single-valued field $\varphi : \partial U_{\mathbf{K}} \to E$, $\varphi(x) = x - a$ obviously do not admit opposite directions and hence, applying Theorem 2.2.17 and the normalization property we get

$$deg_{\mathbf{K}}(\Phi, \partial U_{\mathbf{K}}) = deg_{\mathbf{K}}(\varphi, \partial U_{\mathbf{K}}) = 1.$$

∎

As a corollary we can obtain the following fixed point result for multimaps satisfying *the Leray–Schauder boundary condition*.

Corollary 2.2.25. *Let $0 \in U_{\mathbf{K}}$, $\partial U_{\mathbf{K}} \neq \emptyset$ and $F : \overline{U}_{\mathbf{K}} \to Kv(\mathbf{K})$ a compact multimap such that in each point $x \in \partial U_{\mathbf{K}}$ we have*

$$F(x) \cap \gamma x = \emptyset, \quad \forall \gamma \geq 1.$$

Then $\deg_{\mathbf{K}}(i - F, \partial U_{\mathbf{K}}) = 1$ and hence $Fix\,F \neq \emptyset$.

One more corollary is presented by the fixed point principle for multimaps which are "directed inward" on the boundary (**Prove it**).

Corollary 2.2.26. *Suppose that $U_{\mathbf{K}}$ is convex, $0 \in U_{\mathbf{K}}$, $\partial U_{\mathbf{K}} \neq \emptyset$. Let $F : \overline{U}_{\mathbf{K}} \to Kv(\mathbf{K})$ be a compact multimap having no fixed points on $\partial U_{\mathbf{K}}$ and such that*

$$F(x) \subset I_{\overline{U}_{\mathbf{K}}}(x) = \{y \mid y \in E, \ x + \gamma(y - x) \in \overline{U}_{\mathbf{K}} \ \text{for some } \gamma > 0\}$$

at each point $x \in \partial U_{\mathbf{K}}$. Then $\deg_{\mathbf{K}}(i - F, \partial U_{\mathbf{K}}) = 1$ and hence $FixF \neq \emptyset$.

Consider now *the odd field theorem.*

Theorem 2.2.27. *Suppose that $\mathbf{K}$ and $U_{\mathbf{K}}$ are symmetric with respect to the origin, $0 \in U_{\mathbf{K}}$ and $\partial U_{\mathbf{K}} \neq \emptyset$. Let $F : \overline{U}_{\mathbf{K}} \to Kv(\mathbf{K})$ be a compact multimap such that the corresponding multifield $\Phi = i - F$ satisfies the boundary condition*

$$\Phi(x) \cap \mu\Phi(-x) = \emptyset$$

for all $x \in \partial U_{\mathbf{K}}$ and $0 \leq \mu \leq 1$. Then the topological degree $deg_{\mathbf{K}}(\Phi, \partial U_{\mathbf{K}})$ is an odd number and hence $FixF \neq \emptyset$.

Proof. We start with the showing that for each regular ε-approximation f of the restriction of F to $\partial U_{\mathbf{K}}$ with a sufficiently small $\varepsilon > 0$, the corresponding field $\varphi = i - f$ satisfies the same boundary condition

$$\varphi(x) \neq \mu\varphi(-x)$$

for all $x \in \partial U_{\mathbf{K}}$ and $0 \leq \mu \leq 1$. In fact, supposing the contrary, we will have a sequence of regular ε_n-approximations f_n, $\varepsilon_n \to 0$ such that $\varphi_n(x_n) = \mu_n\varphi_n(-x_n)$ for some $x_n \in \partial U_{\mathbf{K}}$, $\mu_n \in [0,1]$, where $\varphi_n = i - f_n$. This means that

$$x_n = \frac{1}{1+\mu_n}\left[f_n(x_n) - \mu_n f_n(-x_n)\right].$$

Since the sequences $\{f_n(x_n)\}$, $\{f_n(-x_n)\}$ are contained in a relatively compact set $coF(\partial U_{\mathbf{K}})$ we can assume without loss of generality that $f_n(x_n) \to y_0$, $f_n(-x_n) \to z_0$. We set also that $\mu_n \to \mu_0 \in [0,1]$. Therefore $x_n \to x_0 = \frac{1}{1+\mu_0}(y_0 - \mu_0 z_0)$ and so $x_0 - y_0 = \mu_0(-x_0 - z_0)$.

But the pairs $(x_n, f_n(x_n))$, $(-x_n, f_n(-x_n))$ lie in the ε_n-neighborhood of the graph of the restriction of F to $\partial U_{\mathbf{K}}$ and since this graph is closed (Theorem 1.2.29) we have $y_0 \in F(x_0)$, $z_0 \in F(-x_0)$. But this means that $\Phi(x_0) \cap \mu_0\Phi(-x_0) \neq \emptyset$, contrary to the assumption.

The homotopy generated by the family of nondegenerated fields $\psi : \partial U_{\mathbf{K}} \times [0,1] \to E$,

$$\psi(x, \lambda) = \frac{1}{1+\lambda}\left[\varphi(x) - \lambda\varphi(-x)\right],$$

connects the field $\varphi(\cdot) = \psi(\cdot, 0)$ with the odd field $\psi_1 = \psi(\cdot, 1) = \frac{1}{2}[\varphi(x) - \varphi(-x)]$. Considering the field φ as a single-valued homotopic approximation of the multifield Φ and applying Theorem 2.2.5 we obtain that

$$deg_{\mathbf{K}}(\Phi, \partial U_{\mathbf{K}}) = deg_{\mathbf{K}}(\varphi, \partial U_{\mathbf{K}}) = deg_{\mathbf{K}}(\psi_1, \partial U_{\mathbf{K}})$$

is an odd number, proving the theorem. ∎

Notice that as the development of the odd field theorem, the statements on the topological degree of *equivariant* (i.e., commuting with periodic linear operators) multifields can be considered (see, for example, [217], [235]).

To formulate the next statement consider the iterations of a multimap $F : X \to P(E)$, $X \subset E$. Set $F^0 := id_X$, $F^j(x) = F(F^{j-1}(x))$, $j \geq 1$.

Let us prove now a Browder type asymptotic fixed point theorem (F.E. Browder [85]).[1]

Theorem 2.2.28. *Let* $U_{0\mathbf{K}}$, $U_{1\mathbf{K}}$, $U_{\mathbf{K}}$ *be nonempty relatively open subsets of* $\mathbf{K}$, $\overline{U}_{0\mathbf{K}} \subset U_{1\mathbf{K}} \subset \overline{U}_{1\mathbf{K}} \subset U_{\mathbf{K}}$; $U_{0\mathbf{K}}$ *and* $U_{\mathbf{K}}$ *be convex. Suppose that a compact multimap* $F : U_{\mathbf{K}} \to Kv(\mathbf{K})$ *is such that for some integer* $m \geq 1$ *the following conditions hold true:*

(1) $\displaystyle \bigcup_{1 \leq j \leq m-1} F^j(\overline{U}_{1\mathbf{K}}) \subset U_{\mathbf{K}}$;

(2) $\displaystyle \bigcup_{1 \leq j \leq m-1} F^j(\overline{U}_{0\mathbf{K}}) \subset U_{1\mathbf{K}}$;

(3) $F^m(\overline{U}_{1\mathbf{K}}) \subset U_{0\mathbf{K}}$.

Then the multimap F *has a fixed point in the set* $U_{0\mathbf{K}}$.

Proof. Passing, if necessary, to the set $\widetilde{\mathbf{K}} \subset \mathbf{K}$, $\widetilde{\mathbf{K}} = \overline{co}(F(U_{\mathbf{K}}) \cup x_0)$, where $x_0 \in U_{0\mathbf{K}}$ is an arbitrary point, we can assume without loss of generality that the set $\mathbf{K}$ is compact.

The idea of the proof is to find an ε-approximation of the multimap F so fine that it will satisfy all conditions (1) – (3).

Let us consider condition (1). By virtue of the compactness of all sets $F^j(\overline{U}_{1\mathbf{K}})$, $1 \leq j \leq m-1$ and upper semicontinuity of the multimap F we may construct the sequence of numbers

$$\varepsilon_{m-1} > \varepsilon_{m-2} > \ldots > \varepsilon_1 > 0$$

in the following way.

Take $\eta_{m-1} > 0$ such that

$$V_{\eta_{m-1}}(F^{m-1}(\overline{U}_{1\mathbf{K}})) \subset U_{\mathbf{K}},$$

where V_η denotes the η-neighborhood of a set in a relative metric topology of the space $\mathbf{K}$ and set

$$\varepsilon_{m-1} = \frac{1}{2}\eta_{m-1}.$$

---

[1] A fixed point theorem is said to be *asymptotic* if the existence of a fixed point of a map is deduced from the properties of its iterations.

Further, take $\eta_{m-2} > 0$ such that

$$V_{\eta_{m-2}}(F^{m-2}(\overline{U}_{1\mathbf{K}})) \subset U_{\mathbf{K}}$$

and

$$F(V_{\eta_{m-2}}(F^{m-2}(\overline{U}_{1\mathbf{K}}))) \subset V_{\varepsilon_{m-1}}(F^{m-1}(\overline{U}_{\mathbf{K}})).$$

We can assume, without loss of generality that $\eta_{m-2} < \eta_{m-1}$.

Now take $\varepsilon_{m-2} = \frac{1}{3}\eta_{m-2}$ and choose η_{m-3}, $0 < \eta_{m-3} < \eta_{m-2}$ such that

$$V_{\eta_{m-3}}(F^{m-3}(\overline{U}_{1\mathbf{K}})) \subset U_{\mathbf{K}}$$

and

$$F(V_{\eta_{m-3}}(F^{m-3}(\overline{U}_{1\mathbf{K}}))) \subset V_{\varepsilon_{m-2}}(F^{m-2}(\overline{U}_{\mathbf{K}})).$$

Take $\varepsilon_{m-3} = \frac{1}{3}\eta_{m-3}$.

Continuing this process, we construct the decreasing sequence of positive numbers $\varepsilon_j = \frac{1}{3}\eta_j$, $2 \leq j \leq m-2$ such that

$$V_{\eta_j}(F^j(\overline{U}_{1\mathbf{K}})) \subset U_{\mathbf{K}}$$

and

$$F(V_{\eta_j}(F^j(\overline{U}_{1\mathbf{K}}))) \subset V_{\varepsilon_{j+1}}(F^{j+1}(\overline{U}_{\mathbf{K}})). \tag{2.6}$$

Finally, find $0 < \eta_1 < \eta_2$ such that

$$V_{\eta_1}(F(\overline{U}_{1\mathbf{K}})) \subset U_{\mathbf{K}}$$

and

$$F(V_{\eta_1}(F(\overline{U}_{1\mathbf{K}}))) \subset V_{\varepsilon_2}(F^2(\overline{U}_{\mathbf{K}})) \tag{2.7}$$

and take $\varepsilon_1 < \min\{\frac{1}{2}\eta_1, \varepsilon_2\}$.

Now take $\delta' > 0$ so small that the restriction f to $\overline{U}_{1\mathbf{K}}$ of a δ'-approximation $f_{\delta'}: \overline{U}_{\mathbf{K}} \to \mathbf{K}$ of the multimap F should be the ε_1-approximation of the restriction $F|_{\partial U_{1\mathbf{K}}}$ and defines the topological degree of the corresponding multifield $i - F$ on $\partial U_{1\mathbf{K}}$.

Then from relations (2.6) and (2.7) we get

$$f(\overline{U}_{1\mathbf{K}}) \subset V_{\varepsilon_1}(F(\overline{U}_{1\mathbf{K}})) \subset U_{\mathbf{K}};$$

$$f^j(\overline{U}_{1\mathbf{K}}) \subset V_{2\varepsilon_j}(F^j(\overline{U}_{1\mathbf{K}})) \subset V_{\eta_j}(F^j(\overline{U}_{1\mathbf{K}})) \subset U_{\mathbf{K}}, \quad 2 \leq j \leq m-2;$$

$$f^{m-1}(\overline{U}_{1\mathbf{K}}) \subset V_{2\varepsilon_{m-1}}(F^{m-1}(\overline{U}_{1\mathbf{K}})) = V_{\eta_{m-1}}(F^{m-1}(\overline{U}_{1\mathbf{K}})) \subset U_{\mathbf{K}},$$

implying that condition (1) is fulfilled for the map f.

Similarly, we can find numbers $\delta'' > 0$ and $\delta''' > 0$ such that conditions (2) and (3) will be satisfied for the corresponding approximations $f_{\delta''}$ and $f_{\delta'''}$ of the multimap F. Then, if we will take $\delta = \min\{\delta', \delta'', \delta'''\}$ for the δ-approximation f_δ of F all conditions (1) - (3) will be fulfilled.

Now, according to the extension of Yu.G. Borisovich of the Browder theorem [64] from the fulfilment of conditions (1) – (3) for the map f it follows that

$$deg_{\mathbf{K}} (i - f, \partial U_{1\mathbf{K}}) = 1,$$

but then also

$$deg_{\mathbf{K}} (i - F, \partial U_{1\mathbf{K}}) = 1,$$

that implies the existence of a fixed point of F in $U_{1\mathbf{K}}$. Condition (3) yields that all such points belong to $U_{0\mathbf{K}}$. ∎

2.3 Topological degree of condensing multivalued vector fields

> When you jump for joy, beware that no one moves the ground from beneath your feet.
>
> —*Stanislaw Jerzy Lec*

Our aim now is to extend the topological degree theory described in the previous section to a more wide class of multivalued maps.

Let us recall some notions (see, for example, [381], [235]).

Let E be a Banach space, $(\mathcal{A}, \geq)$ a partially ordered set.

Definition 2.3.1. *A map* $\beta : P(E) \to \mathcal{A}$ *is called a measure of noncompactness (MNC) in* E *if*

$$\beta(\overline{co}\,\Omega) = \beta(\Omega) \quad \text{for every } \Omega \in P(E).$$

A MNC β is called:

(i) *monotone if* $\Omega_0, \Omega_1 \in P(E)$, $\Omega_0 \subseteq \Omega_1$ *implies* $\beta(\Omega_0) \leq \beta(\Omega_1)$;
(ii) *nonsingular if* $\beta(\{a\} \cup \Omega) = \beta(\Omega)$ *for every* $a \in E$, $\Omega \in P(E)$;
(iii) *invariant with respect to union with compact sets if* $\beta(K \cup \Omega) = \beta(\Omega)$ *for every relatively compact set* $K \subset E$ *and* $\Omega \in P(E)$;
(iv) *semiadditive if* $\beta(\Omega_0 \cup \Omega_1) = \max\{\beta(\Omega_0), \beta(\Omega_1)\}$ *for every* $\Omega_0, \Omega_1 \in P(E)$;

(v) *real* if $\mathcal{A} = \overline{\mathbb{R}_+} = [0, +\infty]$ with the natural ordering and $\beta(\Omega) < +\infty$ for every bounded set $\Omega \in P(E)$.

If $\mathcal{A}$ is a cone in a Banach space we will say that the MNC β is

(vi) *algebraically semiadditive* if $\beta(\Omega_0 + \Omega_1) \leq \beta(\omega_0) + \beta(\Omega_1)$ for every $\Omega_0, \Omega_1 \in P(E)$;

(vi) *regular* if $\beta(\Omega) = 0$ is equivalent to the relative compactness of Ω.

Well known examples of MNC satisfying all the above properties are:

the Hausdorff MNC

$$\chi(\Omega) = \inf\{\varepsilon > 0 : \Omega \text{ has a finite } \varepsilon\text{-net}\}$$

and *the Kuratowski MNC*

$$\alpha(\Omega) = \inf\{\delta > 0 : \Omega \text{ has a finite partition with sets of}$$
$$\text{diameter less than } \delta\}.$$

We can consider also an example of a MNC defined on bounded subsets of the space $C([a, b]; \mathcal{E})$, where $\mathcal{E}$ is a separable Banach space. For a bounded $\Omega \subset C([a, b]; \mathcal{E})$ set

$$\nu(\Omega) = \big(\sup_{t \in [a,b]} \chi(\Omega(t)), mod_C(\Omega)\big),$$

where χ is the Hausdorff MNC in $\mathcal{E}$, $\Omega(t) = \{d(t) : d \in \Omega\}$ and

$$mod_C(\Omega) = \lim_{\delta \to 0} \sup_{d \in \Omega} \max_{|t_1 - t_2| \leq \delta} \|d(t_1) - d(t_2)\|$$

is the modulus of equicontinuity of the set Ω. The range for the MNC ν is the cone $\mathbb{R}_+^2$ with the natural partial order.

It is worth noting that each component of ν is a real MNC in the same space (see [235]) but, according to the Arzela–Ascoli theorem, only ν possesses the regularity property.

Now, let X be a closed subset of E, β a MNC in E.

Definition 2.3.2. A u.s.c. multimap $F \colon X \to K(E)$ or a u.s.c. family of multimaps $G \colon X \times [0, 1] \to K(E)$ is called β-*condensing* provided for every $\Omega \subset X$ that is not relatively compact we have, respectively,

$$\beta(F(\Omega)) \not\geq \beta(\Omega).$$

or

$$\beta(G(\Omega \times [0, 1])) \not\geq \beta(\Omega).$$

Let us indicate the following important class of condensing multimaps.

Definition 2.3.3. Let β be a real MNC and $0 \le k < 1$. A u.s.c. multimap $F\colon X \to K(E)$ or a u.s.c. family of multimaps $G\colon X \times [0,1] \to K(E)$ is called (k,β)-*condensing* if for every $\Omega \subset X$ we have, respectively,

$$\beta\left(F\left(\Omega\right)\right) \le k\,\beta\left(\Omega\right).$$

or

$$\beta\left(G\left(\Omega \times [0,1]\right)\right) \le k\,\beta\left(\Omega\right).$$

The corresponding multifields of families of multifields are also called $\beta-$ or (k,β)-*condensing*. We will also say about *condensing* multimaps or multifields when the corresponding MNC is implicitly meant.

Consider some examples of condensing multimaps. First of all, from Definition 2.3.3 it follows that every compact multimap is $(0,\chi)$-condensing, where χ is the Hausdorff MNC. A more interesting example is provided by the following assertion.

Theorem 2.3.4. *Let $X \subset E$ be a closed bounded set and a multimap $\Psi\colon X \times E \to K(E)$ satisfy the following conditions:*

(i) *for each $x \in X$ the multimap $\Psi(x,\cdot)\colon E \to K(E)$ is k-Lipschitz with respect to the Hausdorff metric h on $K(E)$, i.e.,*

$$h(\Psi(x,y'),\Psi(x,y'')) \le k\,\|y' - y''\|$$

for every $y',y'' \in E$, where $k \ge 0$ does not depend on x;

(ii) *the set $\Psi(X,y)$ is relatively compact in E for each $y \in E$.*

Then the multimap $\mathcal{F}\colon X \to K(E)$ defined as $\mathcal{F}(x) = \Psi(x,x)$ is (k,χ)-bounded, where χ is the Hausdorff MNC in E, i.e.,

$$\chi(\mathcal{F}(\Omega)) \le k\,\chi(\Omega) \quad \text{for each } \Omega \subset X$$

and hence $\mathcal{F}$ is (k,χ)-condensing if Φ is u.s.c. and $0 \le k < 1$.

Proof. Take $\varepsilon > 0$ and let S be a finite $(\chi(\Omega)+\varepsilon)$-net of Ω in E. We shall show that the relatively compact set $\Psi(\Omega \times S)$ is a $k\,(\chi(\Omega)+\varepsilon)$-net for the set $\mathcal{F}(\Omega)$.

Let $z \in \mathcal{F}(\Omega)$, i.e., $z \in \Psi(x,x)$ with $x \in \Omega$. Take $y \in S$ such that $\|x - y\| \le \chi(\Omega) + \varepsilon$. From condition (i) it follows that there exists a point $z_1 \in \Psi(x,y) \subset \Psi(\Omega \times S)$ such that $\|z - z_1\| \le k\|x - y\| \le k\,(\chi(\Omega)+\varepsilon)$.

Thus $\chi(\mathcal{F}(\Omega)) \le k\,(\chi(\Omega)+\varepsilon)$, and the assertion follows from the arbitrariness of ε. ∎

Corollary 2.3.5. *Let $X \subset E$ be a closed bounded set. If a multimap $F_0 \colon X \to K(E)$ is k-Lipschitz, $0 \le k < 1$, with respect to the Hausdorff metric and a multimap $F_0 \colon X \to K(E)$ is compact then their sum $F_0 + F_1 \colon X \to K(E)$ is (k, χ)-condensing.*

Let us introduce now the following important notion.

Definition 2.3.6. A closed convex subset $\mathcal{K} \subset E$ is called *fundamental* for a multimap $F \colon X \to K(E)$ (or for its corresponding multifield $\Phi = i - F$) if

(1) $F(X \cap \mathcal{K}) \subseteq \mathcal{K}$;
(2) $x \in \overline{co}\,(F\,(x) \cup \mathcal{K})$ implies $x \in \mathcal{K}$.

We emphasize that this definition does not exclude the cases $\mathcal{K} = \emptyset$ or $X \cap \mathcal{K} = \emptyset$ which necessary imply that $Fix\,F = \emptyset$.

Definition 2.3.7. A closed convex subset $\mathcal{K} \subset E$ is called fundamental for a family of multimaps $G \colon X \times [0, 1] \to K(E)$ if it is fundamental for every multimap $G(\cdot, \lambda)$, $\lambda \in [0, 1]$.

The entire space E and $\overline{co}F(X)$ are natural examples of fundamental sets.

The following properties of fundamental sets can be easily verified (**do it!**).

Lemma 2.3.8. *The fixed point set $Fix\,F$ is contained in each fundamental set of F.*

Lemma 2.3.9. *If $\mathcal{K}$ is a fundamental set for a family of multimaps $G \colon X \times [0, 1] \to K(E)$ and $Q \subset \mathcal{K}$ then the set*

$$\widetilde{\mathcal{K}} = \overline{co}(G((X \cap \mathcal{K}) \times [0, 1]) \cup Q)$$

is also fundamental.

Lemma 2.3.10. *If $\{\mathcal{K}_\alpha\}$ is a system of fundamental sets of a multimap F then the set $\cap_\alpha \mathcal{K}_\alpha$ is also fundamental.*

Definition 2.3.11. A fundamental set $\mathcal{K}$ of a multimap $F \colon X \to K(E)$ or a family of multimaps $G \colon X \times [0, 1] \to K(E)$ is said to be *essential* if $X \cap \mathcal{K} \ne \emptyset$ and the restriction of F to $X \cap \mathcal{K}$ or, respectively G to $(X \cap \mathcal{K}) \times [0, 1]$ is compact.

The following assertion is crucial for the construction of the topological degree for a condensing multifield.

Theorem 2.3.12. *Let a family of multimaps* $G\colon X \times [0,1] \to K(E)$ *be* β-*condensing with respect to a monotone nonsingular MNC* β. *Then* G *possesses an essential fundamental set* $\mathcal{K}$.

Proof. Take any point $a \in X$ and consider the collection $\{\mathcal{K}_\alpha\}$ of all fundamental sets of G containing a. Notice that the whole space E belongs to it.

Take the set $\mathcal{K} = \cap_\alpha \mathcal{K}_\alpha$. From Lemmas 2.3.10 and 2.3.9 it follows that $\mathcal{K} \in \{\mathcal{K}_\alpha\}$ and

$$\mathcal{K} = \overline{co}(G((X \cap \mathcal{K}) \times [0,1]) \cup \{a\}). \tag{2.8}$$

Then we have

$$\beta(X \cap \mathcal{K}) \leq \beta(\mathcal{K}) = \beta(\overline{co}(G((X \cap \mathcal{K}) \times [0,1]) \cup \{a\}))$$

$$= \beta((G((X \cap \mathcal{K}) \times [0,1]) \cup \{a\})) = \beta((G((X \cap \mathcal{K}) \times [0,1])))$$

implying that the nonempty set $X \cap \mathcal{K}$ is compact and now the assertion follows from Theorem 1.2.35. ∎

Remark 2.3.13. Notice that from (2.8) and the Mazur theorem (see Chapter 0) it follows that the set $\mathcal{K}$ may be taken compact.

Now we are in position to introduce the notion of the topological degree for a condensing multifield. Let $U \subset E$ be an open set.

Definition 2.3.14. Let a multimap $F\colon \overline{U} \to Kv(E)$ be β-condensing with respect to a monotone nonsingular MNC β and $Fix\, F \cap \partial U = \emptyset$. The *topological degree of the multifield* $\Phi = i - F$ is defined as

$$deg(\Phi, \overline{U}) := deg_\mathcal{K}(\Phi, \partial U_\mathcal{K}),$$

where $\mathcal{K}$ is an arbitrary essential fundamental set of F.

Lemma 2.3.15. *Definition 2.3.14 is consistent, that is the topological degree does not depend on the choice of an essential fundamental set* $\mathcal{K}$.

Proof. Let $\mathcal{K}_0$, $\mathcal{K}_1$ be two essential fundamental sets of F. Then by Lemma 2.3.10 the set $\mathcal{K} = \mathcal{K}_0 \cap \mathcal{K}_1$ is also fundamental. If $\overline{U}_\mathcal{K} = \emptyset$ then from Lemma 2.3.8 it follows that $Fix\, F = \emptyset$ and hence

$$deg_{\mathcal{K}_0}(\Phi, \partial U_{\mathcal{K}_0}) = deg_{\mathcal{K}_1}(\Phi, \partial U_{\mathcal{K}_1}) = 0.$$

In case $\overline{U}_{\mathcal{K}} \neq \emptyset$ we prove that

$$deg_{\mathcal{K}_i}(\Phi, \partial U_{\mathcal{K}_i}) = deg_{\mathcal{K}}(\Phi, \partial U_{\mathcal{K}}), \quad i = 0, 1.$$

Consider a retraction $r \colon E \to \overline{co}(F(\overline{U}_{\mathcal{K}}))$ and the compact multimap $\widehat{F} \colon \overline{U}_{\mathcal{K}_0} \to Kv(\mathcal{K})$, $\widehat{F}(x) = \overline{co}(r \circ F)(x)$. The family $G \colon \overline{U}_{\mathcal{K}_0} \times [0, 1] \to Kv(\mathcal{K}_0)$ defined as

$$G(x, \lambda) = (1 - \lambda)F(x) + \lambda \widehat{F}(x)$$

generates a homotopy $\Phi \sim \widehat{\Phi}$ on $\partial U_{\mathcal{K}_0}$, where $\widehat{\Phi} = i - \widehat{F}$.

In fact, it is clear that the family G is compact. Now, let $x_0 \in G(x_0, \lambda_0)$ for some $x_0 \in \partial U_{\mathcal{K}_0}$, $\lambda_0 \in [0, 1]$. Then $x_0 \in \overline{co}(F(x_0) \cup \mathcal{K})$, hence $x_0 \in \mathcal{K}$. Therefore $G(x_0, \lambda_0) = F(x_0)$ and $x_0 \in Fix\,F$ contrary to the assumption that $Fix\,F \cap \partial U = \emptyset$.

From the homotopy property of the degree of compact multifields (Theorem 2.2.15) it follows that

$$deg_{\mathcal{K}_0}(\Phi, \partial U_{\mathcal{K}_0}) = deg_{\mathcal{K}_0}(\widehat{\Phi}, \partial U_{\mathcal{K}_0}),$$

but the map restriction property (Theorem 2.2.19) implies that

$$deg_{\mathcal{K}_0}(\widehat{\Phi}, \partial U_{\mathcal{K}_0}) = deg_{\mathcal{K}}(\widehat{\Phi}, \partial U_{\mathcal{K}}) = deg_{\mathcal{K}}(\Phi, \partial U_{\mathcal{K}}).$$

Therefore

$$deg_{\mathcal{K}_0}(\Phi, \partial U_{\mathcal{K}_0}) = deg_{\mathcal{K}}(\Phi, \partial U_{\mathcal{K}}).$$

The equality $deg_{\mathcal{K}_1}(\Phi, \partial U_{\mathcal{K}_1}) = deg_{\mathcal{K}}(\Phi, \partial U_{\mathcal{K}})$ can be proved analogously. ∎

Remark 2.3.16. We use the notation $deg(\Phi, \overline{U})$ since in our construction we employ an essential fundamental set $\mathcal{K}$ of the multimap F defined on the whole domain $\overline{U}$. It can be shown that in fact under our conditions the degree does not depend on values taking by the multimap F "inside" the domain, i.e., on U (see [235]).

Definition 2.3.14 provides the possibility to transfer all main properties of the topological degree of compact multifields to the case of condensing ones. Consider, as example, the homotopy invariance property.

Definition 2.3.17. The multifields $\Phi_0 = i - F_0$, $\Phi_1 = i - F_1$ are said to be β-*homotopic*, $\Phi_0 \overset{\beta}{\sim} \Phi_1$, if there exists a β-condensing family of multimaps $G \colon \overline{U} \times [0, 1] \to Kv(E)$ such that $Fix\,G(x, \lambda) = \emptyset$ for all $x \in \partial U$, $\lambda \in [0, 1]$ and $G(\cdot, 0) = F_0$, $G(\cdot, 1) = F_1$.

Theorem 2.3.18. *If* $\Phi_0 \overset{\beta}{\sim} \Phi_1$ *then*

$$deg(\Phi_0, \overline{U}) = deg(\Phi_1, \overline{U}).$$

Proof. From Theorem 2.3.12 we know that there exists an essential fundamental set for the whole family G and then the assertion follows from Theorem 2.2.15. ∎

Obviously, we have the following general fixed point principle.

Theorem 2.3.19. *Let* $F: \overline{U} \to Kv(E)$ *be a* β-*condensing multimap which is fixed point free on* ∂U. *If* $deg(i - F, \overline{U}) \neq 0$, *then* $\emptyset \neq Fix\, F \subset U$.

It is clear also that all fixed point results considered in the previous section for compact multimaps have their "condensing" analogs (see [235]). We distinguish only the following "multivalued" version of the Darbo–Sadovskii fixed point theorem which is often used in applications. This result generalizes at the same time the Bohnenblust–Karlin theorem (Theorem 2.2.22).

Theorem 2.3.20. *If* M *is a convex closed subset of* E *and* $F: M \to Kv(M)$ *is a* β-*condensing multimap, where* β *is a nonsingular MNC, then* $Fix\, F \neq \emptyset$.

Proof. Since M is convex we can assume, without loss of generality, that $M \cap \mathcal{K} = \mathcal{K}$ for any fundamental set $\mathcal{K}$ of F. But then the monotonicity property of β in the proof of the existence of an essential fundamental set of F (see Theorem 2.3.12) is superfluous. Take a restriction of F to such a set and apply Theorem 2.2.22. ∎

Further properties of condensing multimaps and their applications to the study of differential inclusions in Banach spaces can be found in monograph [235].

2.4 Some properties of the fixed point set

> If it is hard for you to understand all infinity at once, try to understand it at least halfway.
>
> *—Slawomir Wróblewski*

In this section we describe some properties of the fixed point set of compact multimaps which are important for forthcoming applications.

Theorem 2.4.1. *Let* X *be a closed subset of a Banach space* E *and* $F: X \to K(E)$ *a closed multimap such that the image* $F(\Omega)$ *of every bounded* set $\Omega \subset X$ *is relatively compact. If the fixed point set* $FixF$ *is bounded then it is compact.*

Proof. As the multimap F is closed, it follows that the set $Fix\, F$ is closed. Then its compactness follows from the relation

$$FixF \subset F\,(FixF)\,.$$

∎

Consider now the question on the continuous dependence of the fixed point set on a parameter.

Theorem 2.4.2. *Let X be a closed subset of a Banach space E; Λ a metric space and $G : X \times \Lambda \to K\,(E)$ a closed multimap such that the set $G\,(X \times \Delta)$ is relatively compact for each compact set $\Delta \subset \Lambda$. Suppose that*

$$\mathcal{F}\,(\lambda) := FixG\,(\cdot, \lambda) \neq \emptyset$$

for all $\lambda \in \Lambda$. Then the multimap $\mathcal{F} : \Lambda \to K\,(E)$ is upper semicontinuous.

Proof. From conditions of the theorem it easily follows that each set $\mathcal{F}(\lambda)$, $\lambda \in \Lambda$ is compact. Assuming that the multimap $\mathcal{F}$ is not u.s.c. we may find a point $\lambda_0 \in \Lambda$ and sequences $\{\lambda_n\} \subset \Lambda$, $\lambda_n \to \lambda_0$, $\{x_n\} \subset X$ such that

$$x_n \in \mathcal{F}\,(\lambda_n) \setminus V$$

for all n, where V is a certain neighborhood of the set $\mathcal{F}\,(\lambda_0)$. Since $x_n \in G\,(x_n, \lambda_n)$ for all n and the sequence $\{\lambda_n\}$ is relatively compact, from the condition of the theorem we obtain that the sequence $\{x_n\}$ is also relatively compact and hence we may assume, without loss of generality, that $x_n \to x_0 \in X$. But then the closedness of the multimap G implies $x_0 \in G\,(x_0, \lambda_0)$, i.e., $x_0 \in \mathcal{F}\,(\lambda_0)$, giving the contradiction. ∎

To verify that the fixed point sets $\mathcal{F}\,(\lambda)$ are nonempty we can apply the fixed point principles described in Section 2.2. A convenient opportunity can be provided also by the following stability property of the topological degree.

Theorem 2.4.3. *Let Λ be a metric space; $G : \partial U_{\mathbf{K}} \times \Lambda \to Kv\,(\mathbf{K})$ a upper semicontinuous multimap such that the set $G\,(\partial U_{\mathbf{K}} \times \Delta)$ is relatively compact for each compact set $\Delta \subset \Lambda$. Suppose that for some $\lambda_0 \in \Lambda$ we have $Fix\, G\,(\cdot, \lambda_0) = \emptyset$. Then for all $\lambda \in \Lambda$ sufficiently close to λ_0 the degree*

$$deg_{\mathbf{K}}\,(i - G\,(\cdot, \lambda)\,, \partial U_{\mathbf{K}})$$

is well defined and coincides with $deg_{\mathbf{K}}\,(i - G\,(\cdot, \lambda_0)\,, \partial U_{\mathbf{K}})\,.$

Proof. For all $\lambda \in \Lambda$ sufficiently close to λ_0, the multimap $\mathcal{G}_\lambda : \partial U_{\mathbf{K}} \times [0,1] \to Kv(\mathbf{K})$,

$$\mathcal{G}_\lambda(x,\mu) = \mu G(x,\lambda) + (1-\mu)G(x,\lambda_0)$$

defines the homotopy connecting the compact multifields $i - G(\cdot,\lambda_0)$ and $i - G(\cdot,\lambda)$. In fact, supposing the contrary we have sequences $\{\lambda_n\} \subset \Lambda$, $\lambda_n \to \lambda_0$, $\{\mu_n\} \subset [0,1]$, $\mu_n \to \mu_0 \in [0,1]$ and $\{x_n\} \subset \partial U_{\mathbf{K}}$ such that

$$x_n \in \mathcal{G}_{\lambda_n}(x_n,\mu_n)$$

for all n.

The sequence $\{x_n\}$ is contained in the relatively compact set

$$co\,G\left(\{x_n\} \times \overline{\{\lambda_n\}}\right)$$

so we can assume, without loss of generality, that $x_n \to x_0 \in \partial U_{\mathbf{K}}$. It is easy to verify (see Section 1.3) that the multimap $\mathcal{G}_\lambda(x,\mu)$ is u.s.c. and hence closed with respect to the variables λ, x and μ. Then, passing to the limit, we get

$$x_0 \in \mathcal{G}_{\lambda_0}(x_0,\mu_0) = G(x_0,\lambda_0),$$

contrary to the condition of the theorem. It remains to apply the homotopy invariance property (Theorem 2.2.15). ∎

Corollary 2.4.4. *Let Λ be a metric space; $G : \overline{U}_{\mathbf{K}} \times \Lambda \to Kv(\mathbf{K})$ a upper semicontinuous multimap such that the set $G\left(\overline{U}_{\mathbf{K}} \times \Delta\right)$ is relatively compact for each compact set $\Delta \subset \Lambda$. Suppose that for some $\lambda_0 \in \Lambda$ we have $Fix\,G(\cdot,\lambda_0) \cap \partial U_{\mathbf{K}} = \emptyset$ and $deg_{\mathbf{K}}(i - G(\cdot,\lambda_0), \partial U_{\mathbf{K}}) \neq 0$. Then for all $\lambda \in \Lambda$ sufficiently close to λ_0, the multimap*

$$\mathcal{F}(\lambda) = Fix\,G(\cdot,\lambda)$$

is well defined and is upper semicontinuous.

Remark 2.4.5. Analogous properties hold true also for condensing multimaps (see [235], Section 3.5).

2.5 The Browder–Ky Fan fixed point theorem and variational inequalities

> The weakest link in a chain is the strongest because it can break it.
>
> —*Stanislaw Jerzy Lec*

At the present time, the fixed point theory is extended, besides the above considered classes of multivalued maps, to many other ones. Consider one of examples. A refined result, given below (see [86], [133]) is interesting by the fact that it is valid in an arbitrary linear topological space and the closedness of values for the multimap under consideration is not required.

Theorem 2.5.1. *Let M be a nonempty convex compact subset of a linear topological space X and a multimap $F : M \to Pv(M)$ be such that the preimages $F_-^{-1}(y)$ are relatively open subsets of M for every $y \in M$. Then $Fix\, F \neq \emptyset$.*

Proof. By the compactness of M there exists a finite set of points $y_1, ..., y_n$ of M such that the sets $F_-^{-1}(y_i)$, $i = 1, ..., n$ form the cover of M. Since a compact space is paracompact, let $p_1, ..., p_n$ be a partition of unity subordinated to this cover (see Chapter 0).

Let $M_0 = \overline{co}\{y_1, ..., y_n\}$. Then M_0 is a convex compact subset of a finite-dimensional space $X_0 = sp\{y_1, ..., y_n\}$. Consider a continuous map $f : M_0 \to M_0$ given as

$$f(x) = \sum_{i=1}^{n} p_i(x) y_i.$$

It is easy to see that it is a selection of F on the set M_0. In fact, if $p_i(x) \neq 0$ for some $x \in M_0$ then this means that $x \in F_-^{-1}(y_i)$ or $y_i \in F(x)$ and hence $f(x) \in F(x)$ by virtue of the convexity of the set $F(x)$. According to the Brouwer theorem, the map f has a fixed point which obviously is also a fixed point of F.   ∎

The proved statement may be used for the solving of variational inequalities. Indeed, consider the following problem.

Let X be a Hilbert space with an inner product $\langle \cdot, \cdot \rangle$; M a convex compact subset of X and $t : M \to X$ a continuous map.

Theorem 2.5.2. *There exists a point $x_0 \in M$ such that*

$$\langle t(x_0), x_0 - x \rangle \geq 0 \ \text{for all} \ \ x \in M.$$

Proof. Suppose the contrary. Then for every point $x_0 \in M$ there exists a point $x \in M$ such that

$$\langle t(x_0), x_0 - x \rangle < 0.$$

Then define the multimap $F : M \to Pv(M)$ by the formula

$$F(x_0) = \{x \in M \mid \langle t(x_0), x_0 - x \rangle \; < \; 0\}.$$

It is easy to see **(verify it!)** that the multimap F satisfies conditions of Theorem 2.5.1 and hence it possesses a fixed point that leads to the contradiction. ∎

Besides variational inequalities, Theorem 2.5.1 and its numerous extensions find applications in many parts of nonlinear functional analysis (theorems of Knaster–Kuratowski–Mazurkiewicz type), results in theory of games (minimax relations) and mathematical economics (an equilibrium and Pareto optimums) (see, for example, survey [73] and monograph [409]).

Chapter 3

Differential inclusions and control systems

> From a shooter's experience: a big target is more difficult to hit than a small one.
>
> —*Stanislaw Jerzy Lec*

3.1 Differential inclusions. Some examples

> All that I desire to point out is the general principle that life imitates art far more than art imitates life.
>
> —*Oscar Wilde*

Some time ago *differential inclusions* were called (not quite exactly) "differential equations with a multivalued right-hand side". In a fairly general form the differential inclusion can be written in the form of the relation

$$x'(t) \in F(t, x(t)),$$

where F is a certain multimap.

The arising of the theory of differential inclusions dates from the thirties of the last century. The pioneering works of this direction were the researches of the French mathematician A.Marchaud [301]– [304] and a Polish mathematician S.K.Zaremba [410], [411]. In these papers they proved the existence theorems for "differential equations in contingents and paratingents" and described the main properties of solutions. The fate of many discoveries that passed ahead of their hour awaited these works: they were forgotten for some time. Notice however that in forties-fifties the differential inclusions were used in the papers of A.D. Myshkis [313], [314].

In the course of time it became clear that the differential inclusions provide a convenient tool for the description of implicit differential equations and differential inequalities. In fact, relations of the form

$$f(t, x(t), x'(t)) = 0$$

or

$$f(t, x(t), x'(t)) \geq 0$$

can be written in the form of differential inclusions

$$x'(t) \in F(t, x(t)),$$

where the multimap F is defined as

$$F(t, x) = \{z : f(t, x, z) = 0\}$$

or, respectively,

$$F(t, x) = \{z : f(t, x, z) \geq 0\}.$$

Differential inclusions may be applied also to the study of differential equations whose right-hand side is known only up to some degree of exactness or is a subject to systematic perturbations. In this case the corresponding description can be achieved by means of the relation of the form

$$x'(t) \in f(t, x(t)) + \varepsilon B,$$

where B is a certain ball.

However "the hour of triumph" of the theory of differential inclusions has struck on the border-line of sixties when (first of all, in the works of A.F. Filippov [137] and T. Ważewski [403], [404]) a deep and natural connection between differential inclusions and control systems became clear. The interest to control problems in the years after the World War II was caused by urgent needs of technology, aviation, astronautics, power engineering. It is just during this period that there appear such general methods for solving control problems as Pontryagin's maximum principle, Bellman's dynamic programming method and others.

Sufficiently adequate mathematical description of a feedback control system may be reduced to the relations of the form

$$\begin{cases} x'(t) = f(t, x(t), u(t)) & (3.1) \\ u(t) \in U(t, x(t)), & (3.2) \end{cases}$$

where the function f characterizes the dynamics of the system, $x(t)$ is its *trajectory*, and $u(t)$ is *a control function* whose values at each moment t are selected from the corresponding set of admissible controls U depending, in the general case, both on the time t and on the state of the system $x(t)$.

The multimap U is called *the feedback multifunction*. The control function u is realized usually in the class of measurable functions.

Consider, for example, the problem of regulation of the rotation of a satellite around its axis. Let ρ be the moment of inertia of the satellite with respect to the axis, $\omega(t)$ the angular speed of the rotation at the moment t, $u(t)$ the moment of forces created by the control engines with respect to the axis of the rotation. Then the equation describing the rotation of the satellite will have the form

$$\rho\omega'(t) = u(t).$$

While the study of this process, it is necessary to take into account the variation of possibilities of the engines caused, for example, by the pressure drop in the containers with the working matter. Therefore, the control moment is subject to constraints $u(t) \in U(t)$, where U is a certain multifunction.

In more refined evaluations it is necessary to consider also the dependence of the control possibilities on the state of the system that is reflected in the choice of the feedback multifunction.

Let us return to the general case of control system (3.1)–(3.2). It is clear that if a pair $(x(t), u(t))$ defines a trajectory of the system and a corresponding control then this trajectory is a solution of the associated differential inclusion

$$x'(t) \in f(t, x(t), U(t, x(t))). \tag{3.3}$$

However, the inverse transfer from inclusion (3.3) to system (3.1)–(3.2) is not so evident. The equivalence of relations (3.1)–(3.2) and (3.3), i.e., the fact that for each solution x of inclusion (3.3) there exists such a measurable control function u which realizes it as a trajectory of system (3.1)–(3.2) is established with the help of the Filippov implicit function lemma (see Theorem 1.5.15). We will consider this question in Section 3.4.

Approximately at the same time differential inclusions have found one more important and interesting application. It was their employment for the study of systems described by differential equations with discontinuous right-hand side (see [138], [141]). It is known that in the classical theory of differential equations the right-hand side is, as a rule, continuous with respect to the phase variables. If the right-hand side is discontinuous, an equation can have no solutions even in simplest cases. For example, the Cauchy problem

$$x'(t) = \begin{cases} 1, & x < 0, \\ -1, & x \geq 0, \end{cases}$$

$$x\left(0\right) = 0,$$

has no solutions in usual sense. (**Verify it!**).

However differential equations with discontinuous right-hand side very naturally arise in many applications. For example, a large number of problems from mechanics and electrical engineering lead to differential equations with discontinuous right-hand sides (see, e.g., [386]). We will restrict ourselves to the following model of the autopilot.

Let φ be the angle between the desired heading of the aircraft and its actual heading. If this angle is small enough, the angular motion of the aircraft can be described by the differential equation

$$\varphi'' = -k\varphi' + \tau\left(\varphi\right),$$

where τ is a torque created by aircraft's stabilization control according to the following simple principle:

$$\tau\left(\varphi\right) = \begin{cases} T, & \varphi < 0, \\ -T, & \varphi > 0. \end{cases}$$

One of approaches suggested for the investigations of equations of this type was very natural: to replace the discontinuous right-hand part $f\left(x\right)$ with the multivalued one defined by the formula

$$F\left(x\right) = \bigcap_{\varepsilon > 0} \overline{co} f\left(B_\varepsilon\left(x\right)\right).$$

It is easy to see that the operation of such kind does not change a value $f\left(x\right)$ if x is a point of continuity but "dares the holes" at discontinuity points. So, the above mentioned function $\tau\left(\varphi\right)$ after this transformation takes the value $[-T,T]$ at $\varphi = 0$.

It seems natural to call a solution of the differential inclusion

$$x'\left(t\right) \in F\left(x\left(t\right)\right)$$

a generalized solution of an initial differential equation

$$x'\left(t\right) = f\left(x\left(t\right)\right)$$

with a discontinuous right-hand part. This approach turned out to be very fruitful. The point is that in the case of a bounded right-hand side f the multimap F is upper semicontinuous independently on the class of f and this opens the possibilities to apply to the corresponding differential inclusion the technique of existence theorems and other methods (see, for example, Theorem 3.2.15).

Differential inclusions and connected problems arise very naturally not only in the investigation of technical or transport systems. Let us present an example from the sphere of mathematical economics. Consider *a model of economical dynamics of von Neumann–Gale type.*

Let the state of an economics at a given moment is characterized by a certain n-dimensional vector x. The components of x conform to the amounts of the corresponding goods in the system at the present moment. Suppose that for each state of the economics $x \in \mathbb{R}^n$ at a moment t it is given a set $A(x) \subset \mathbb{R}^n$ of states to which the economics may pass by the moment $t+1$. Therefore the given economical model is defined by a certain multimap $A : \mathbb{R}^n \to P(\mathbb{R}^n)$ characterizing the productive and trade capabilities of the system. The sequence of states of the system $\{x(t_0 + k)\}_{k=0}^{\infty}$ at moments $\{t_0 + k\}_{k=0}^{\infty}$ is called *a technologically admissible trajectory* of the system provided

$$x(t+1) \in A(x(t))$$

for all $t = t_0 + k$, $k = 0, 1, ...$

Now, let us make a passage from the given discrete model to a model functioning in a continuous time. Suppose that the process of transformation of goods is going on uniformly (i.e., at the moment $t + \tau$, $\tau \in (0,1)$ the state of the economics may be characterized by any of the vectors of the set $(1 - \tau)x(t) + \tau Ax(t)$. Then

$$\frac{x(t+\tau) - x(t)}{\tau} \in \frac{(1-\tau)x(t) + \tau A(t) - x(t)}{\tau} = Ax(t) - x(t).$$

Passing to the limit as $\tau \to 0$ we get the differential inclusion describing the continuous dynamics of the economical system

$$x'(t) \in A(x(t)) - x(t).$$

3.2 Existence theorems and properties of the solution sets

> Arguments are extremely vulgar, for everyone in good society holds exactly the same opinion.
>
> *—Oscar Wilde*

In this section we will consider the problems of solvability of the Cauchy problem for a differential inclusion in a finite-dimensional space:

$$\begin{cases} x'(t) \in F(t, x(t)) & (3.4) \\ x(t_0) = x_0, & (3.5) \end{cases}$$

where $F : I \times \mathbb{R}^n \to K(\mathbb{R}^n)$ is a multimap, $I = [t_0, T] \subset \mathbb{R}$ an interval endowed with the Lebesgue measure μ.

Problem (3.4)–(3.5) can be studied under various assumptions imposed on the right-hand part F of the inclusion and we can put different meanings in the notion of a solution. We can consider, for example, *a classical solution* supposing that the function x is everywhere differentiable and that inclusion (3.4) holds true at each point of the interval I. However, in the sequel we will consider only *the Carathéodory type solutions*. To be more precise, let us give the following definition.

Definition 3.2.1. *A solution of Cauchy problem* (3.4)–(3.5) *on some interval* $[t_0, \tau]$, $t_0 < \tau \leq T$ *is an absolutely continuous function* $x : [t_0, \tau] \to \mathbb{R}^n$ *satisfying initial condition* (3.5) *and inclusion* (3.4) *at almost each point of the interval* $[t_0, \tau]$.

For finding solutions to the Cauchy problem we will apply the fixed point methods developed in the previous chapter.

Suppose that a multimap F is superpositionally selectable and satisfies the condition of local integrable boundedness (*F*3) (see Definitions 1.5.21 and 1.5.25). Then the integral multioperator (cf. 1.5.33)

$$j \circ \mathcal{P}_F : C([t_0, \tau]; \mathbb{R}^n) \to P(C([t_0, \tau]; \mathbb{R}^n))$$

is well defined.

As the base for our approach we will use the following statement, the verifying of validity of which we leave to the readers.

Theorem 3.2.2. *A function* $x : [t_0, \tau] \to \mathbb{R}^n$ *is a solution to problem* (3.4)–(3.5) *if and only if it is a fixed point of the multioperator*

$$\mathcal{J}_F : C([t_0, \tau]; \mathbb{R}^n) \to P(C([t_0, \tau]; \mathbb{R}^n)),$$

$$\mathcal{J}_F = x_0 + j \circ \mathcal{P}_F.$$

In the sequel the following property of the multioperator $\mathcal{J}_F$ will play an important role.

Lemma 3.2.3. *The image* $\mathcal{J}_F(D) \subset C([t_0, \tau]; \mathbb{R}^n)$ *of every bounded set* $D \subset C([t_0, \tau]; \mathbb{R}^n)$ *is relatively compact.*

Proof. We will use the Arzela–Ascoli theorem (see Chapter 0). Let $y \in \mathcal{J}_F(D)$ be an arbitrary function. This means that y has the form

$$y(t) = x_0 + \int_{t_0}^t f(s)\, ds,$$

where $f \in \mathcal{P}_F(x)$, $x \in D$. Take $r > 0$ such that $\|x\| \le r$, $\forall x \in D$. Then for each $t \in [t_0, \tau]$ we have

$$\|y(t)\| \le \|x_0\| + \int_{t_0}^t \|f(s)\|\, ds \le \|x_0\| + \int_{t_0}^\tau \nu_r(s)\, ds,$$

i.e., the set $\mathcal{J}_F(D)$ is bounded.

On the other hand, for each $t_1, t_2 \in [t_0, \tau]$ we get:

$$\|y(t_2) - y(t_1)\| = \| \int_{t_1}^{t_2} f(s)\, ds \| \le \int_{t_1}^{t_2} \|f(s)\|\, ds \le \int_{t_1}^{t_2} \nu_r(s)\, ds,$$

i.e., the set $\mathcal{J}_F(D)$ is equicontinuous. ∎

We can prove now the following *local existence theorem.*

Theorem 3.2.4. *Let a multimap $F : I \times \mathbb{R}^n \to Kv(\mathbb{R}^n)$ satisfy the conditions:*

   (F1) *for each $x \in \mathbb{R}^n$ the multifunction $F(\cdot, x) : I \to Kv(\mathbb{R}^n)$ admits a measurable selection;*
   (F2) *for μ-a.e. $t \in I$ the multimap $F(t, \cdot) : \mathbb{R}^n \to Kv(\mathbb{R}^n)$ is upper semicontinuous*

and condition of local integral boundedness (F3) (see Definition 1.5.25).

Then there exists τ, $t_0 < \tau \le T$ such that Cauchy problem (3.4)–(3.5) has a solution on the interval $[t_0, \tau]$.

Proof. Take an arbitrary number $R > 0$ and let, for a number $r = \|x_0\| + R$ the function ν_r correspond to r in accordance with condition (F3). Choose now τ, $t_0 < \tau \le T$ sufficiently close to t_0 to provide

$$\int_{t_0}^\tau \nu_r(s)\, ds \le R.$$

Then for the ball $\mathcal{B} \subset C([t_0, \tau]; \mathbb{R}^n)$,

$$\mathcal{B} = \{y \in C([t_0, \tau]; \mathbb{R}^n) \mid \|y - \overline{x}_0\|_C \le R\}$$

with $\overline{x}_0(t) \equiv x_0$, we obviously have

$$\mathcal{J}_F(\mathcal{B}) \subset \mathcal{B}.$$

From Corollary 1.5.34 it follows that the multioperator $\mathcal{J}_F : \mathcal{B} \to Cv(\mathcal{B})$ is closed, whereas Lemma 3.2.3 and Theorem 1.2.32 yield its compactness. From the Bohnenblust–Karlin theorem (Theorem 2.2.22) we deduce that the multimap $\mathcal{J}_F$ has a fixed point and the application of Theorem 3.2.2 concludes the proof.   ∎

Further we will need a slightly modified version of an assertion about integral inequalities which is known under the name of *the Gronwall lemma* (see, for example, [204], Theorem III.1.1; [121], Remark 1.7.86)

Lemma 3.2.5. *Let $u, v : [a, b] \to \mathbb{R}$ be nonnegative functions, u be integrable and v continuous; $C \geq 0$ a constant and*

$$v(t) \leq C + \int_a^t u(s)\, v(s)\, ds, \quad a \leq t \leq b.$$

Then

$$v(t) \leq C e^{\int_a^t u(s)ds}, \qquad a \leq t \leq b.$$

Replace now condition $(F3)$ of local integral boundedness of the right-hand part with a more strong *condition of sublinear growth*

$(F3')$ there exists a function $\alpha \in L^1_+(I)$ such that

$$\|F(t, x)\| \leq \alpha(t)(1 + \|x\|) \quad \text{for a.e. } t \in I$$

for all $x \in \mathbb{R}^n$.

We can formulate now *the global existence theorem.*

Theorem 3.2.6. *Let a multimap $F : I \times \mathbb{R}^n \to Kv(\mathbb{R}^n)$ satisfy conditions $(F1)$, $(F2)$, $(F3')$. Then Cauchy problem (3.4)-(3.5) has a solution on the interval $[t_0, T]$.*

For the proof we need the following statement.

Lemma 3.2.7. *Under conditions of Theorem 3.2.6 the solution set of an one-parameter family of Cauchy problems*

$$\begin{cases} x'(t) \in \lambda F(t, x(t)), & t \in I, \lambda \in [0, 1] & (3.6) \\ x(t_0) = x_0, & (3.7) \end{cases}$$

is a priori bounded.

Proof. Each function being a solution to problem (3.6)-(3.7) for a certain $\lambda \in [0,1]$ has the form

$$x(t) = x_0 + \lambda \int_{t_0}^{t} f(s)\, ds,$$

where $f \in \mathcal{P}_F(x)$. But then for a continuous function $v(t) = \|x(t)\|$ we have the following estimate:

$$v(t) \le \|x_0\| + \lambda \int_{t_0}^{t} \|f(s)\|\, ds \le \|x_0\| + \int_{t_0}^{t} \alpha(s)(1 + \|x(s)\|)\, ds \le$$

$$\le \|x_0\| + \int_{t_0}^{T} \alpha(s)\, ds + \int_{t_0}^{t} \alpha(s) v(s)\, ds.$$

Applying Lemma 3.2.5 we get

$$\|x(t)\| \le C e^{\int_{t_0}^{T} \alpha(s)ds}, \quad t \in [t_0, T],$$

where $C = \|x_0\| + \int_{t_0}^{T} \alpha(s)\, ds$. ∎

Proof of Theorem 3.2.6. Consider the convex closed set $\mathbf{K} \subset C(I; \mathbb{R}^n)$,

$$\mathbf{K} = \{ y \in C(I; \mathbb{R}^n) \mid y(t_0) = x_0 \}.$$

By Lemma 3.2.7 we can indicate a relatively open bounded subset $U_{\mathbf{K}}$ containing all solutions to problem (3.6)-(3.7). (Notice that this set contains also the function $x_0(t) \equiv x_0$ since it is a solution for $\lambda = 0$).

It is easy to see that the multimap

$$\mathcal{G}_F : \overline{U}_{\mathbf{K}} \times [0,1] \to Kv(\mathbf{K}),$$

$$\mathcal{G}_F(x, \lambda) = x_0 + \lambda j \circ \mathcal{P}_F(x)$$

is compact and, moreover, $x \notin \mathcal{G}_F(x, \lambda)$ for all $(x, \lambda) \in \partial U_{\mathbf{K}} \times [0,1]$ since otherwise such a function x should be a solution to problem (3.6)-(3.7). But this means that the family $\mathcal{G}_F$ defines the homotopy of the multifields $i - \mathcal{G}_F(\cdot, 1) = i - \mathcal{J}_F$ and $i - \mathcal{G}_F(\cdot, 0) = i - x_0$ on $\partial U_{\mathbf{K}}$.

Applying now the homotopy invariance and normalization properties of the topological degree (Theorems 2.2.15 and 2.2.14) we get

$$deg_{\mathbf{K}}(i - \mathcal{J}_F, \partial U_{\mathbf{K}}) = deg_{\mathbf{K}}(i - x_0, \partial U_{\mathbf{K}}) = 1.$$

We conclude the proof with the application of the general fixed point principle (Theorem 2.2.20) and Theorem 3.2.2. ∎

The proved theorem can be used for the solving of the question about the extension of a solution to the Cauchy problem.

Theorem 3.2.8. *Under conditions of Theorem 3.2.6 every solution to Cauchy problem (3.4), (3.5) on an interval $[t_0, \tau]$, $t_0 < \tau < T$ can be extended to a solution on the whole interval $[t_0, T]$.*

Proof. Let $\widetilde{x}$ be a solution to problem (3.4), (3.5) on an interval $[t_0, \tau]$. Consider the Cauchy problem

$$\begin{cases} x'(t) \in F(t, x(t)) \\ x(\tau) = \widetilde{x}(\tau). \end{cases}$$

By Theorem 3.2.6 this problem has a solution $\overline{x}$ on the interval $[\tau, T]$. Joining the solutions $\widetilde{x}$ and $\overline{x}$ we get the desired extension. ∎

In the sequel the set of all solutions to Cauchy problem (3.4)–(3.5) will be denoted by $\Sigma(x_0)$.

The applied operator method allows to obtain some topological properties of the solution set to Cauchy problem as direct consequences of corresponding properties of the fixed point set. The following property plays an important role in the solving of optimization problems for control systems governed by differential inclusions (see Section 3.4.)

Theorem 3.2.9. *Under conditions of Theorem 3.2.6 the set $\Sigma(x_0) \subset C(I; \mathbb{R}^n)$ is compact.*

Proof. The assertion follows from Theorems 3.2.2, 2.3.1 and Lemmas 3.2.3 and 3.2.7. ∎

By using Example 1.2.22 (b) we get the following statement.

Corollary 3.2.10. *Under conditions of Theorem 3.2.6 the multifunction $\Pi : I \to K(\mathbb{R}^n)$,*

$$\Pi(t) = \Sigma(x_0)(t) = \{x(t) \mid x \in \Sigma(x_0)\}$$

is continuous.

Remark 3.2.11. It can be shown that under conditions of Theorem 3.2.6 the solution set $\Sigma(x_0)$ is connected and hence also all the sets $\Pi(t)$, $t \in I$ are connected (see, for example, [2], [112], [116], [190], [392]). Concerning more refined topological properties of the set $\Sigma(x_0)$ see the section "Bibliographical comments and additions".

The following assertion on the continuous dependence of the solution set on initial data holds true.

Theorem 3.2.12. *Under conditions of Theorem 3.2.6 the multimap $\Sigma : \mathbb{R}^n \to K(C(I; \mathbb{R}^n))$ is upper semicontinuous.*

Proof. If we will consider the multioperator $\mathcal{J}_F$ as a multimap of two variables: a function $x \in C(I; \mathbb{R}^n)$ and a parameter $x_0 \in \mathbb{R}^n$ then it is easy to see that this multimap satisfies conditions of Theorem 2.3.2. ∎

We can use also Theorem 2.3.2 for the study of a question on the dependence of the solution set to Cauchy problem on a parameter.

Let Λ be a compact set and a multimap $F : I \times \mathbb{R}^n \times \Lambda \to Kv(\mathbb{R}^n)$ satisfy the following conditions:

$F1_\lambda$) for every $x \in \mathbb{R}^n$, $\lambda \in \Lambda$ the multifunction $F(\cdot, x, \lambda) : I \to Kv(\mathbb{R}^n)$ admits a measurable selection;

$F2_\lambda$) for μ-a.e. $t \in I$ the multimap $F(t, \cdot, \cdot) : \mathbb{R}^n \times \Lambda \to Kv(\mathbb{R}^n)$ is upper semicontinuous;

$F3_\lambda$) there exists a function $\alpha \in L^1_+(I)$ such that
$$\|F(t, x, \lambda)\| \le \alpha(t)(1 + \|x\|) \quad \text{for a.e. } t \in I$$
for all $x \in \mathbb{R}^n$, $\lambda \in \Lambda$.

For each fixed $\lambda \in \Lambda$ we can define the multioperator
$$\mathcal{J}_{F_\lambda} = x_0 + j \circ \mathcal{P}_{F_\lambda},$$
where $F_\lambda(t, x) = F(t, x, \lambda)$.

By using the methods of previous reasonings we can easily verify the following assertion.

Theorem 3.2.13. *For every bounded subset $\Omega \subset C(I; \mathbb{R}^n)$ the family of multioperators*
$$\mathcal{G} : \Omega \times \Lambda \to Kv(C(I; \mathbb{R}^n)), \quad \mathcal{G}(\cdot, \lambda) = \mathcal{J}_{F_\lambda}$$
is compact.

Consider now the Cauchy problem for an one-parametric family of differential inclusions:
$$\begin{cases} x'(t) \in F(t, x(t), \lambda), & t \in I, \lambda \in \Lambda \\ x(t_0) = x_0, \end{cases} \quad \begin{matrix} (3.8) \\ (3.9) \end{matrix}$$

By using Theorems 2.3.2, 3.2.6 and 3.2.13, we come to the following conclusion.

Theorem 3.2.14. *Under conditions $(F1_\lambda)$, $(F2_\lambda)$ and $(F3'_\lambda)$ one-parametric Cauchy problem (3.8)–(3.9) has a nonempty set of solutions $\Sigma_\lambda(x_0)$ for each $\lambda \in \Lambda$ and the multimap $\Xi : \Lambda \to K(C(I; \mathbb{R}^n))$,*
$$\Xi(\lambda) = \Sigma_\lambda(x_0)$$
is upper semicontinuous.

Consider now (not striving to the maximal generality) as the application of obtained results the question about the solvability of the Cauchy problem for a differential equation with a discontinuous right-hand side. The following statement holds true.

Theorem 3.2.15. *Let a map* $f : \mathbb{R}^n \to \mathbb{R}^n$ *be bounded on bounded subsets of* $\mathbb{R}^n$. *Then the Cauchy problem*

$$\begin{cases} x'(t) = f(x(t)), & (3.10) \\ x(t_0) = x_0, & (3.11) \end{cases}$$

has a generalized solution on a certain interval $[t_0, \tau], \ \tau > t_0$.

Proof. Recall (see Section 3.1) that by a generalized solution of problem (3.10)–(3.11) we call a solution of the Cauchy problem for the differential inclusion

$$\begin{cases} x'(t) \in F(x(t)), & (3.12) \\ x(t_0) = x_0, & (3.13) \end{cases}$$

where

$$F(x) = \bigcap_{\varepsilon > 0} \overline{co} f(B_\varepsilon(x)).$$

It is easy to see that the multimap F has nonempty convex compact values at each point $x \in \mathbb{R}^n$ and is also bounded on bounded subsets.

So, to prove the upper semicontinuity of F it is sufficient to show that it is closed. Consider sequences $x_k \to x_*$, $y_k \in F(x_k)$, $y_k \to y_*$.

Without loss of generality we will assume that the sequence

$$\varepsilon_k = 2\|x_k - x_*\| \to 0$$

is decreasing.

Since

$$y_k \in F(x_k) = \bigcap_{\varepsilon > 0} \overline{co} f(B_\varepsilon(x_k))$$

we get

$$y_k \in \overline{co} f(B_\varepsilon(x_k))$$

for every $\varepsilon > 0$. For a fixed k take now $\varepsilon > 0$ so small that

$$B_\varepsilon(x_k) \subset B_{\varepsilon_k}(x_*).$$

Then

$$y_k \in \overline{co} f(B_{\varepsilon_k}(x_*)) \quad k = 1, 2, \dots$$

Since the balls $B_{\varepsilon_k}(x_*)$ $k = 1, 2, \ldots$ are embedded into each other we obtain that, if we fix a certain $k = m$ then

$$y_k \in \overline{co} f\left(B_{\varepsilon_m}(x_*)\right)$$

for all $k \geq m$. But then also

$$y_* \in \overline{co} f\left(B_{\varepsilon_m}(x_*)\right).$$

Since m is arbitrary, this means that

$$y_* \in F(x_*).$$

Now we may apply Theorem 3.2.4 to problem (3.12)–(3.13). ∎

By using the same method and Theorem 3.2.6 we may obtain the existence of a generalized solution to problem (3.10)–(3.11) on an arbitrary interval $[t_0, T]$ in the case when the map f is globally bounded.

Consider now the question about the solvability of Cauchy problem (3.4)–(3.5) in the case when the right-hand part F is almost lower semicontinuous (see Definition 1.5.35) and satisfies condition of local integral boundedness $(F3)$. As we know from Theorem 1.5.36 in this case the superposition multioperator $\mathcal{P}_F : C(I; \mathbb{R}^n) \to C(L^1(I; \mathbb{R}^n))$ is lower semicontinuous. Moreover, $\mathcal{P}_F$ obviously has decomposable values (see Definition 1.4.7). Then, by Theorem 1.4.8 the multioperator $\mathcal{P}_F$ admits a continuous selection

$$\rho_F : C(I; \mathbb{R}^n) \to L^1(I; \mathbb{R}^n).$$

But then the continuous operator $\iota_F : C(I; \mathbb{R}^n) \to P(C(I; \mathbb{R}^n))$,

$$\iota_F = x_0 + j \circ \rho_F$$

is in turn a continuous selection of the multioperator $\mathcal{J}_F$ and its fixed points are the fixed points of this multioperator. It opens the opportunity in the above operator approach to replace the multioperator $\mathcal{J}_F$ with its continuous selection ι_F and to apply to it the corresponding "single-valued" fixed point principles (the theorem of Schauder and the degree of compact fields). As the result we get the following existence theorems, whose proofs we left to the reader.

Theorem 3.2.16. *Let a multimap* $F : I \times \mathbb{R}^n \to K(\mathbb{R}^n)$ *be almost lower semicontinuous and satisfy condition* $(F3)$. *Then there exists* τ, $t_0 < \tau \leq T$, *such that Cauchy problem (3.7)-(3.5) has a solution on the interval* $[t_0, \tau]$.

Theorem 3.2.17. *Let a multimap $F : I \times \mathbb{R}^n \to K\left(\mathbb{R}^n\right)$ be almost lower semicontinuous and satisfy condition $(F3')$. Then Cauchy problem (3.4)-(3.5) has a solution on interval I.*

Concerning topological properties of the solution set in the lower semicontinuous case let us mention the following assertion.

Theorem 3.2.18. *Under conditions of Theorem 3.2.17 the solution set $\Sigma\left(x_0\right)$ is connected and hence the sets*

$$\Pi\left(t\right) = \Sigma\left(x_0\right)\left(t\right), \ t \in I$$

are also connected.

The proof of this result can be found, for example, in [116], [235].

3.3 Periodic solutions of differential inclusions

> Suppose you succeed in breaking the wall with your head. And what, then, will you do in the next prison cell?
>
> —*Stanislaw Jerzy Lec*

The problem of searching the periodic solutions is among classical ones in the theory of differential equations. In the present section we will apply the above developed topological methods to the study of this problem for differential inclusions.

Let a multimap $F : \mathbb{R} \times \mathbb{R}^n \to Kv\left(\mathbb{R}^n\right)$ be T-periodic in t, i.e.,

$$F\left(t,x\right) = F\left(t+T,x\right), \ T > 0$$

for all $t \in \mathbb{R}$, $x \in \mathbb{R}^n$. It is clear that this condition opens the opportunity to consider the multimap F only on the domain $[0,T] \times \mathbb{R}^n$. We will assume that the multimap F satisfies on this set conditions $(F1)$, $(F2)$, $(F3)$ of the previous section.

A solution x of a differential inclusion

$$x'\left(t\right) \in F\left(t, x\left(t\right)\right) \tag{3.14}$$

satisfying the boundary value condition of periodicity

$$x\left(0\right) = x\left(T\right). \tag{3.15}$$

is called *a T-periodic solution.* It is clear that every such function can be extended to a T-periodic solution defined on the whole real axis $\mathbb{R}$.

Our nearest purpose is the reducing of the question about the existence of a periodic solution to the fixed point problem for a multivalued integral operator in a functional space. This will allow to apply the topological methods developed in Chapter 2 to theorems on the existence of periodic solutions.

Denote $\mathcal{C} = C\left([0,T];\mathbb{R}^n\right)$. Apparently, the simplest integral multioperator intended for the search of T-periodic solutions is the multioperator $\mathcal{J}_T : \mathcal{C} \to Kv\left(\mathcal{C}\right)$ of the form

$$\mathcal{J}_T\left(x\right) = x\left(T\right) + j \circ \mathcal{P}_F\left(x\right).$$

(See Definition 1.5.3).

The following assertion can be verified directly **(do it)!**.

Theorem 3.3.1. *Fixed points of the multioperator $\mathcal{J}_T$ coincide with solutions to periodic problem* (3.14)–(3.15).

Similarly to how it was done for the integral multioperator $\mathcal{J}_F$ (see Lemma 3.2.3) we can verify that the image $\mathcal{J}_T\left(D\right)$ of each bounded set $D \subset \mathcal{C}$ is relatively compact. Further, by using Corollary 1.5.34 it is easy to check that the multioperator $\mathcal{J}_T$ is closed and hence upper semicontinuous (Theorem 1.2.32).

Therefore, the topological degree theory developed in Chapter 2 is applicable to the multioperator $\mathcal{J}_T$ and we can formulate the following general principle.

Theorem 3.3.2. *Let $\mathbf{K}$ be a closed convex set in the space $\mathcal{C}$; $U_{\mathbf{K}} \subset \mathbf{K}$ a bounded relatively open subset. If $deg_{\mathbf{K}}\left(i - \mathcal{J}_T, \partial U_{\mathbf{K}}\right) \neq 0$ then periodic problem* (3.14)–(3.15) *has a solution in $U_{\mathbf{K}}$.*

Consider some simple theorems on the existence of periodic solutions of differential inclusions.

Let a multimap $F : \mathbb{R} \times \mathbb{R}^n \times [0,1] \to Kv\left(\mathbb{R}^n\right)$ be T-periodic in the first argument and its restriction to $[0,T] \times \mathbb{R}^n \times [0,1]$ satisfy conditions $(F1_\lambda)$–$(F2_\lambda)$ of the previous section and the following parametric condition of local integral boundedness

$(F3_\lambda)$ for each $r > 0$ there exists a function $\nu_r \in L^1_+\left(I\right)$ such that

$$\|F\left(t,x,\lambda\right)\| \leq \nu_r\left(t\right) \quad \text{for a.e. } t \in I$$

for all $x \in \mathbb{R}^n$, $\|x\| \leq r$, $\lambda \in [0,1]$.

Then we can consider the one-parametric family of differential inclusions

$$x'\left(t\right) \in F\left(t, x\left(t\right), \lambda\right), \quad \lambda \in [0,1] \tag{3.16}$$

which generates the family of integral multioperators $\mathcal{G}_T : \mathcal{C} \times [0,1] \to Kv(\mathcal{C})$,

$$\mathcal{G}_T(x, \lambda) = x(T) + j \circ \mathcal{P}_{F_\lambda}(x),$$

where $F_\lambda = F(\cdot, \cdot, \lambda)$.

It is easy to see that if $D \subset \mathcal{C}$ is an arbitrary bounded set then the restriction of $\mathcal{G}_T$ to $D \times [0,1]$ is compact and hence the family $\mathcal{G}_T$ can generate the homotopy of integral multioperators.

Let $U_{\mathbf{K}} \subset \mathbf{K}$ be a bounded relatively open subset such that $\partial U_{\mathbf{K}}$ does not contain T-periodic solutions of family (3.16) for all $\lambda \in [0,1]$ (suppose, for example, that all T-periodic solutions of this family belonging to $\mathbf{K}$ admit a priori estimate by virtue of which they all lie in $U_{\mathbf{K}}$). Then, as a direct consequence of the homotopy invariance principle (Theorem 2.2.15) and Theorem 3.3.2 we get the following statement.

Theorem 3.3.3. *Let* $deg_{\mathbf{K}}(i - \mathcal{G}_T(\cdot, 0), \partial U_{\mathbf{K}}) \neq 0$. *Then the periodic problem for the differential inclusion*

$$x'(t) \in F(t, x(t), 1)$$

has a solution in $U_{\mathbf{K}}$.

In particular, the following results hold true.

Corollary 3.3.4. *Let* U *is convex and* $\mathcal{G}_T(x, 0) \cap \overline{U}_{\mathbf{K}} \neq \varnothing$ *for all* $x \in \partial U_{\mathbf{K}}$. *Then the conclusion of Theorem 3.3.3 holds true.*

Proof. From the proof of Theorem 2.2.21 it follows that

$$deg_{\mathbf{K}}(i - \mathcal{G}_T(\cdot, 0), \partial U_{\mathbf{K}}) = 1.$$

∎

Corollary 3.3.5. *Let* $\mathbf{K}$ *and* $U_{\mathbf{K}}$ *are symmetric with respect to the origin and* $0 \in U_{\mathbf{K}}$. *Suppose that* $F(t, -x, 0) = -F(t, x, 0)$ *for all* $x \in \mathbb{R}^n$ *and almost all* $t \in [0, T]$. *Then the conclusion of Theorem 3.3.3 holds true.*

Proof. It is easy to see that the integral multioperator $\mathcal{G}_T(\cdot, 0)$ corresponding to the right-hand part $F(\cdot, \cdot, 0)$ is odd:

$$\mathcal{G}_T(-x, 0) = -\mathcal{G}_T(x, 0)$$

and it remains to apply the odd field theorem (Theorem 2.2.27).   ∎

In the course of the study of some concrete periodic problems certain integral multioperators of a form different from $\mathcal{J}_T$ might appear to be useful. In particular, let us consider the case of a differential inclusion with a separated linear part or *a semilinear differential inclusion*

$$x'(t) \in ax(t) + F(t, x(t)), \tag{3.17}$$

where $a : \mathbb{R}^n \to \mathbb{R}^n$ is a linear operator and a multimap F is T-periodic and satisfies the above conditions $(F1)$, $(F2)$ and $(F3)$.

We will suppose that the operator a satisfies the condition: *1 does not belong to the spectrum of the operator* e^{aT} and hence the linear operator $\left[i - e^{aT}\right]^{-1}$ is well defined.

Define the multioperator $\mathcal{I}_T : \mathcal{C} \to P(\mathcal{C})$ in the following way:

$$\mathcal{I}_T(x) = \left\{ y \mid y(t) = e^{at} \left[i - e^{aT}\right]^{-1} \int_0^T e^{a(T-s)} f(s) \, ds + \right.$$

$$\left. + \int_0^t e^{a(t-s)} f(s) \, ds, \quad f \in \mathcal{P}_F(x) \right\}.$$

Theorem 3.3.6. *The set of all solutions to periodic problem* (3.17), (3.15) *coincides with the fixed point set* $Fix\mathcal{I}_T$.

Proof. 1) It is easy to see that each solution x of inclusion (3.17) can be written in the form

$$x(t) = e^{at} x(0) + \int_0^t e^{a(t-s)} f(s) \, ds,$$

where $f \in \mathcal{P}_F(x)$. If this solution satisfies condition of T-periodicity (3.15) then it satisfies

$$x(0) = e^{aT} x(0) + \int_0^T e^{a(T-s)} f(s) \, ds,$$

and hence

$$x(0) = \left[i - e^{aT}\right]^{-1} \int_0^T e^{a(T-s)} f(s) \, ds,$$

implying

$$x(t) = e^{at} \left[i - e^{aT}\right]^{-1} \int_0^T e^{a(T-s)} f(s) \, ds + \int_0^t e^{a(t-s)} f(s) \, ds,$$

i.e., $x \in \mathcal{I}_T(x)$.

2) Now, let a function x be a fixed point of $\mathcal{I}_T$. Then

$$x(t) = e^{at} \left[i - e^{aT}\right]^{-1} \int_0^T e^{a(T-s)} f(s) \, ds + \int_0^t e^{a(t-s)} f(s) \, ds,$$

where $f \in \mathcal{P}_F(x)$. From here it follows that x is a solution of inclusion (3.17). Further,

$$x(T) = e^{aT} \left[i - e^{aT}\right]^{-1} \int_0^T e^{a(T-s)} f(s)\, ds + \int_0^T e^{a(T-s)} f(s)\, ds$$

$$= \left(e^{aT} \left[i - e^{aT}\right]^{-1} + i\right) \int_0^T e^{a(T-s)} f(s)\, ds$$

$$= \left[i - e^{aT}\right]^{-1} \int_0^T e^{a(T-s)} f(s)\, ds = x(0),$$

i.e., x satisfies periodicity condition (3.15).   ∎

Applying Theorem 1.5.30 and the above developed methods for the investigation of integral multioperators we can easily verify that the multioperator $\mathcal{I}_T$ has convex compact values and is compact on bounded subsets of the space $\mathcal{C}$. This circumstances also open the opportunities for the applications of topological methods. As example, let us consider the following results.

Theorem 3.3.7. *Let there exists an a priori estimate for all T-periodic solutions of a family of inclusions*

$$x'(t) \in ax(t) + \lambda F(t, x(t)), \quad \lambda \in [0, 1]. \tag{3.18}$$

Then the set of all solutions to periodic problem (3.17), (3.15) *is nonempty and compact.*

Proof. If $\mathcal{I}_T$ is an integral multioperator for problem (3.17), (3.15) then for each $\lambda \in [0, 1]$ the multimap $\lambda \mathcal{I}_T$ is the integral multioperator for the corresponding inclusion from family (3.18). Then, for a sufficiently large $r > 0$ and the ball $B_r(0) \subset \mathcal{C}$ the family of multifields $i - \lambda \mathcal{I}_T$ generates the homotopy of the identical field i and the multifield $i - \mathcal{I}_T$ on $\partial B_r(0)$. By using the properties of normalization and homotopy invariance of the topological degree we come to the conclusion that

$$\deg\left(i - \mathcal{I}_T, \partial B_r(0)\right) = 1$$

and hence the fixed point set $Fix\mathcal{I}_T$ is nonempty and contained in $B_r(0)$. Its compactness follows from Theorem 2.3.1.   ∎

Corollary 3.3.8. *Let the multivalued part F of inclusion* (3.17) *admits the global estimate*

$(F3'')$ *there exists a function* $\gamma \in L^1_+[0,T]$ *such that*

$$\|F(t,x)\| \leq \gamma(t) \quad \text{for a.e. } t \in [0,T]$$

for all $x \in \mathbb{R}^n$.

Then the conclusion of Theorem 3.3.7 holds true.

Proof. It is sufficient to notice that the fixed points x of the multimaps from the family $\lambda \mathcal{I}_T$, $\lambda \in [0,1]$ in this case admit the a priori estimate

$$\|x\| \leq M\|\gamma\|_{L^1} \left(M\| [i - e^{aT}]^{-1} \| + 1 \right),$$

where $M = \sup_{t \in [0,T]} \|e^{at}\|$. ∎

One more effective tool for the proof of the results on the existence of periodic solutions is *the method of guiding functions*. The base of this method for the case of differential equations was laid in the works of M.A. Krasnoselskii, A.I. Perov and their pupils (see, for example, [260], [262]).

For simplicity, we will restrict ourselves to the case when the right-hand part of a differential inclusion is upper semicontinuous.

Let us introduce the following notion.

Definition 3.3.9. A point $x_0 \in R^n$ is called *a T-non-recurrence point* of trajectories of differential inclusion (3.14) if for each solution x emanating from x_0 the following condition holds:

$$x(t) \neq x_0, \quad \forall t \in (0,T]. \tag{3.19}$$

The following statement plays the key role in the justification of the method of guiding functions.

Theorem 3.3.10. *Let* $U \subset \mathbb{R}^n$ *be a bounded open set such that each point* $x \in \partial U$ *is a T-non-recurrence point of trajectories of the inclusion. Let a multimap* $F : [0,T] \times \mathbb{R}^n \to Kv(\mathbb{R}^n)$ *be upper semicontinuous and satisfy condition* $(F3')$. *If the multifield* $R_0 : \overline{U} \to Kv(\mathbb{R}^n)$,

$$R_0(x) = -F(0,x),$$

does not have zero points on ∂U *then*

$$deg(\Phi, \partial\Omega) = deg(R_0, \partial U),$$

where $\Phi = i - \mathcal{J}_T$ *is the multifield generated by the integral multioperator* $\mathcal{J}_T$ *and* Ω *is a certain bounded open set in the space* $\mathcal{C}$.

Proof. From Lemma 3.2.7 it follows that the set of all solutions to inclusion (3.14) emanating from $\overline{U}$ is bounded. Let $m > 0$ be a number such that the norm of every solution from this set is less than m. Define an open set Ω in the space $\mathcal{C}$ by the following conditions:

$$\Omega = \{x \in \mathcal{C} \mid x(0) \in U, \ \|x\| < m\}.$$

Consider the family of multimaps

$$F_\lambda(t, x) = F(\lambda t, x), \ \lambda \in [0, 1]$$

and the family of multifields $\Psi : \overline{\Omega} \times [0, 1] \to Kv(\mathcal{C})$ defined in the following way:

$$\Psi(x, \lambda) = \Big\{z \mid z(t) = x(t) - x(T) - \lambda \int_0^t f(s)\,ds - (1 - \lambda) \int_0^T f(s)\,ds,$$

$$f \in \mathcal{P}_{F_\lambda}(x)\Big\}.$$

By using Lemma 3.2.3 it is easy to verify that the family of multifields Ψ is compact. Let us show that this family is nondegenerate on $\partial\Omega \times [0, 1]$. Suppose the contrary, i.e., let there exist a function $x_0 \in \partial\Omega$ and a number $\lambda_0 \in [0, 1]$ such that $0 \in \Psi(x_0, \lambda_0)$. It means that there is an integrable selection $f(s) \in F(\lambda_0 s, x_0(s))$ which satisfies the following equality:

$$x_0(t) = x_0(T) + \lambda_0 \int_0^t f(s)\,ds + (1 - \lambda_0) \int_0^T f(s)\,ds, \qquad (3.20)$$

for each $t \in [0, T]$.

For $t = 0$ we have

$$x_0(0) = x_0(T) + (1 - \lambda_0) \int_0^T f(s)\,ds,$$

while for $t = T$ we get

$$\int_0^T f(s)\,ds = 0. \qquad (3.21)$$

Whence $x_0(0) = x_0(T)$.

By taking the derivative in t in both sides of equality (3.20) we obtain

$$x_0'(t) = \lambda_0 f(t) \in \lambda_0 F(\lambda_0 t, x_0(t))$$

for a.e. $t \in [0, T]$.

So, x_0 is a solution of the differential inclusion

$$x'(t) \in \lambda_0 F(\lambda_0 t, x(t)).$$

Notice that by construction of the set Ω, its boundary consists of functions of the following two types:

(1) $x(0) \in \partial U$;

(2) $x(0) \in U$, $\|x\| = m$.

Consider two cases:

(a) $\lambda_0 = 0$,

(b) $\lambda_0 \neq 0$.

Case (a). If $\lambda_0 = 0$ then $x_0(t) \equiv x_0$ for each $t \in [0, T]$, $f(t) \in F(0, x_0)$ for a.e. $t \in [0, T]$ and from (3.21) we obtain $0 \in F(0, x_0)$.

The function $x_0(t)$, being a constant, can not be a function of the first type since, by condition, the multifield R_0 has no zero points on ∂U.

On the other hand, the function $x_0(t)$ can not be a second type function, since $\|x_0\| < m$ by construction of the set Ω.

Case (b). If now $\lambda_0 \neq 0$, let us consider the function $z_0(t) = x_0\left(\frac{t}{\lambda_0}\right)$. Then for a.e. $t \in [0, \lambda_0 T]$ we have

$$z_0'(t) = \frac{1}{\lambda_0} x_0'\left(\frac{t}{\lambda_0}\right) = f\left(\frac{t}{\lambda_0}\right) \in F\left(t, x_0\left(\frac{t}{\lambda_0}\right)\right) = F(t, z_0(t)).$$

So, the function z_0 is a solution of differential inclusion (3.19) on the interval $[0, \lambda_0 T]$. According to the global existence theorem, we can extend it to the whole interval $[0, T]$.

Notice that the function x_0 can not be the first type function. Indeed, from

$$x_0(0) = z_0(0) = x_0(T) = z_0(\lambda_0 T)$$

it follows that inclusion (3.19) has a solution z_0 such that $z_0(0) \in \partial U$ and $z_0(0) = z_0(\lambda_0 T)$, contrary to the assumption that trajectories emanating from ∂U are T-non-recurrent.

On the other hand, x_0 can not be the second type function. It follows from

$$\|x_0\| \leq \|z_0\| < m$$

since z_0 is a solution of inclusion (3.19) emanating from the set U.

So, the family of multifields Ψ realizes the homotopy of multifields

$$\Psi_1 = \Phi = i - \mathcal{J}_T,$$

and

$$\Psi_0(x) = i - \Gamma_0(x),$$

where the multioperator $\Gamma_0 : \overline{U} \to Kv\,(\mathcal{C})$ is defined by the relation

$$\Gamma_0\,(x) = x\,(T) + \int_0^T F\,(0, x\,(s))\,ds.$$

This multioperator acts into the finite-dimensional subspace $C^n_{[0,T]}$ of constant functions being naturally isomorphic to $\mathbb{R}^n$. By using the map restriction property of the topological degree (Theorem 2.2.19) we obtain

$$deg\,(\Psi_0; \partial U) = deg\,(\Psi_0\,|_{R^n},\, \partial\,(U \cap \mathbb{R}^n))\,.$$

Applying Theorem 1.5.12(e), notice that the multifield $\widehat{\Psi}_0 = \Psi_0\,|_{\mathbb{R}^n}$ is defined by the relations

$$\widehat{\Psi}_0\,(x) = -\int_0^T F\,(0, x)\,ds = -T \cdot F\,(0, x)\,.$$

Finally, we obtain

$$deg\,(\Phi, \partial\Omega) = deg\,(\Psi_0, \partial\Omega) = deg\,(-F\,(0, \cdot)\,, \partial U) = deg\,(R_0, \partial U)\,.$$

∎

The following assertion on the existence of a periodic solution immediately follows from the proved result.

Corollary 3.3.11. *In conditions of Theorem 3.3.10, let*

$$deg\,(R_0, \partial U) \neq 0.$$

Then differential inclusion (3.14) has a T-periodic solution.

Now we are in position to describe the idea of *the method of guiding functions*. Let us introduce the following notions.

A continuously differentiable function $v : \mathbb{R}^n \to \mathbb{R}$ is called *the nondegenerate potential* if its gradient is non-zero outside a certain ball centered at the origin, i.e., there exists $r_v > 0$ such that

$$grad\,v\,(x) = \left\{ \frac{\partial v\,(x)}{\partial x_1}, \frac{\partial v\,(x)}{\partial x_2}, ..., \frac{\partial v\,(x)}{\partial x_n} \right\} \neq 0,$$

for each $x \in \mathbb{R}^n$, $\|x\| \geq r_v$.

From the properties of the topological degree 2.2.3 and 2.2.6 it follows that the degree of the gradient of a nondegenerate potential

$$deg\,(grad\,v\,(x), S_r)$$

on the spheres $S_r \subset \mathbb{R}^n$ of radius $r \geq r_v$ does not depend on r. This generic value of the degree is called *the index of a nondegenerate potential* and it is denoted by $ind\, v$.

As an example of a potential with a non-zero index we can consider a nondegenerate potential v satisfying the coercivity condition

$$\lim_{\|x\| \to \infty} |v(x)| \to \infty. \tag{3.22}$$

(see [262]).

Other examples of potentials with non-zero index can be found in [260], [262].

Definition 3.3.12. A nondegenerate potential v is called *a guiding function* for differential inclusion (3.14) if

$$\langle grad\, v(x), y \rangle > 0$$

for all $0 \leq t \leq T$, $\|x\| \geq r_v$, $y \in F(t, x)$.

From this definition it follows immediately that if v is a guiding function of inclusion (3.14) then the field $-grad\, v$ and the multifield R_0 do not allow opposite directions on spheres S_r of radius $r \geq r_v$ and hence, by Theorem 2.2.17

$$deg(R_0, S_r) = (-1)^n\, ind\, v. \tag{3.23}$$

(We have used the known property of the degree of single-valued fields: $deg(-\varphi, S) = (-1)^n\, deg(\varphi, S)$, see, for example, [260]).

We can formulate now the following condition for the existence of a periodic solution.

Theorem 3.3.13. *Let $F : [0, T] \times \mathbb{R}^n \to Kv(\mathbb{R}^n)$ be a upper semicontinuous multimap satisfying sublinear growth condition $(F3')$. If for differential inclusion (3.14) there exists a guiding function v of a non-zero index then the inclusion has a T-periodic solution.*

Proof. Let $r_0 \geq r_v$ be defined as

$$r_0 = \left(r_v + \int_0^T \alpha(s)\, ds \right) e^{\int_0^T \alpha(s) ds},$$

where α is the function from condition $(F3')$.

Then if x is a solution of inclusion (3.14) with initial condition $\|x(0)\| > r_0$ then $\|x(t)\| > r_v$ for all $t \in [0, T]$. Indeed, suppose to the contrary that there exists $t_0 \in (0, T]$ such that $\|x(t_0)\| \leq r_v$. For $t \in [0, t_0]$ define

$$y(t) = x(t_0 - t), \quad \beta(t) = \alpha(t_0 - t), \quad G(t, x) = -F(t_0 - t, x).$$

It is clear that

$$y'(t) \in G(t, y(t)).$$

Since $\|G(t, x)\| \le \beta(t)(1 + \|x\|)$ for a.e. $t \in [0, t_0]$, applying Lemma 3.2.7 we get

$$\|y(t)\| \le \left(\|y(0)\| + \int_0^{t_0} \beta(s)\,ds \right) e^{\int_0^{t_0} \beta(s)ds} \le r_0$$

for all $t \in [0, t_0]$. Therefore, $\|x(0)\| = \|y(t_0)\| \le r_0$ and we have the contradiction.

Now, if we take an arbitrary r, $r > r_0$ then the sphere S_r consists of T-non-recurrence points of inclusion (3.14). In fact, if x is a solution of (3.14) such that $x(0) \in S_r$ then, as it was shown above, $\|x(t)\| > r_v$ for all $t \in [0, T]$. Then for each $t \in (0, T]$ we have

$$v(x(t)) - v(x(0)) = \int_0^t \langle grad\, v(x(s)), x'(s) \rangle\, ds > 0,$$

implying (3.19).

To conclude the proof, it remains to apply relation (3.23) and Corollary 3.3.11. ∎

Corollary 3.3.14. *Let $F : [0, T] \times \mathbb{R}^n \to Kv(\mathbb{R}^n)$ be a upper semicontinuous multimap satisfying condition $(F3')$. If for differential inclusion (3.14) there exists a guiding function v satisfying coercivity condition (3.22) then the inclusion has a T-periodic solution.*

The method of guiding functions for differential inclusions allows extensions in various directions (see, for example, monographs [190], [339]).

Present here one of the most general results (see [190]).

Theorem 3.3.15. *Let a multimap $F : [0, T] \times \mathbb{R}^n \to Kv(\mathbb{R}^n)$ satisfy conditions $(F1)$, $(F2)$ and $(F3')$. Let there exist a nondegenerate potential $v : \mathbb{R}^n \to \mathbb{R}$ of a non-zero index such that for each $x \in \mathbb{R}^n$, $\|x\| \ge r_v$ and $t \in [0, T]$ there exists $y \in F(t, x)$ such that*

$$\langle grad\, v(x), y \rangle \ge 0.$$

Then inclusion (3.14) has a T-periodic solution.

3.4 Control systems

I have the simplest tastes. I am always satisfied with the best.

—Oscar Wilde

We will consider *a feedback control system* of the form
$$\begin{cases} x'(t) = f(t, x(t), u(t)), & (3.24) \\ u(t) \in U(t, x(t)). & (3.25) \end{cases}$$
Here $f : I \times \mathbb{R}^n \times \mathbb{R}^m \to \mathbb{R}^n$ is *the map characterizing the dynamics of the system*; as earlier, $I = [t_0, T]$; $\mathbb{R}^m$ is *the space of control parameters*; $U : I \times \mathbb{R}^n \to K(\mathbb{R}^m)$ is *the feedback multifunction*.

We will suppose that the following conditions are fulfilled:

($f1$) for each $(x, u) \in \mathbb{R}^n \times \mathbb{R}^m$ the function $f(\cdot, x, u) : I \to \mathbb{R}^n$ is measurable;

($f2$) for a.e. $t \in I$ the map $f(t, \cdot, \cdot) : \mathbb{R}^n \times \mathbb{R}^m \to \mathbb{R}^n$ is continuous;

($U1$) the multimap U satisfies the upper Carathéodory conditions (see Definition 1.5.13);

($U2$) the multimap U is superpositionally measurable (see Theorems 1.5.18 and 1.5.19);

($U3$) the set
$$F(t, x) = f(t, x, U(t, x))$$
is convex for all $(t, x) \in I \times \mathbb{R}^n$;

($U4$) the multimap $F : I \times \mathbb{R}^n \to Kv(\mathbb{R}^n)$ satisfies the sublinear growth condition $(F3')$.

By *a solution of control system* (3.24), (3.25) we mean a pair $\{x, u\}$, consisting of the trajectory x and the control u. Here $x : I \to \mathbb{R}^n$ is an absolutely continuous function satisfying equation (3.24) a.e. on I and $u : I \to \mathbb{R}^m$ is a measurable function satisfying inclusion (3.25) everywhere on I.

From control system (3.24), (3.25) we can pass to the associated differential inclusion
$$x'(t) \in F(t, x(t)). \tag{3.26}$$
As it was mentioned above, each trajectory of system (3.24), (3.25) is a solution of inclusion (3.26). The Filippov implicit function lemma (Theorem 1.5.15) allows to establish the inverse relationship.

Theorem 3.4.1. *Under above conditions, for each solution x of inclusion (3.26) there exists a measurable function $u : I \to \mathbb{R}^m$ such that the pair $\{x, u\}$ is a solution of control system* (3.24), (3.25).

Proof. Let $x : I \to \mathbb{R}^n$ be a solution of inclusion (3.26). Consider the map $\varphi : I \times \mathbb{R}^m \to \mathbb{R}^n$, $\varphi(t, u) = f(t, x(t), u)$. It is easy to see that the map φ satisfies the Carathéodory conditions.

Indeed, the continuity of φ in u for a.e. fixed $t \in I$ directly follows from condition $(f2)$. Further, since for each fixed $u \in \mathbb{R}^m$, the map $f(\cdot, \cdot, u) :$ $I \times \mathbb{R}^n \to \mathbb{R}^n$ satisfies the Carathéodory conditions, it is superpositionally measurable. So, for each fixed $u \in \mathbb{R}^m$ the map $\varphi(\cdot, u) = f(\cdot, x(\cdot), u) :$ $I \to \mathbb{R}^n$ is measurable.

By virtue of condition $(U2)$ the multifunction $V : I \to K(\mathbb{R}^m)$, $V(t) = U(t, x(t))$ is measurable. The function x satisfies the inclusion $x'(t) \in$ $\varphi(t, V(t))$ for a.e. $t \in I$. From Theorem 1.5.15 it follows that there exists a measurable selection $u : I \to \mathbb{R}^m$ of the multifunction V which will satisfy for a.e. $t \in I$ the relation $x'(t) = \varphi(t, u(t))$, i.e., $x'(t) = f(t, x(t), u(t))$.

But this means that the function $u : I \to \mathbb{R}^m$ is the desired control which realizes the solution x of inclusion (3.26) as the trajectory of control system (3.14), (3.25). ∎

The equivalence between control systems and differential inclusions established by Theorem 3.4.1 allows to transfer easily to control systems the above proved statements concerning differential inclusions. Moreover, the investigated properties of the solution sets of differential inclusions can be used for solving optimization problems. Let us consider some examples.

Theorem 3.4.2. *Let $\phi : C(I; \mathbb{R}^n) \to \mathbb{R}$ be a lower semicontinuous functional. Then under conditions $(f1)$, $(f2)$ and $(U1)$-$(U4)$ there exists a solution $\{x_*, u_*\}$ of system (3.14), (3.25) satisfying initial condition*

$$x(t_0) = x_0 \tag{3.27}$$

and such that

$$\phi(x_*) = \min_{x \in \Sigma(x_0)} \phi(x),$$

where $\Sigma(x_0)$ is the set of all trajectories of control system satisfying initial condition (3.27).

Proof. Let us show that the multimap F determining the right-hand side of associated differential inclusion (3.26) satisfies the conditions of the global existence theorem (Theorem 3.2.6). Since the fulfilment of condition $(F3')$ is provided by condition $(U4)$, it remains to prove only conditions $(F1)$ and $(F2)$.

Let $t \in I$ be such that the map $f(t, \cdot, \cdot)$ is continuous and the multimap $U(t, \cdot)$ is u.s.c. Then

$$F(t, \cdot) = f(t, \cdot, U(t, \cdot)) = f(t, \cdot, \cdot) \circ (id_{R^n} \times U(t, \cdot))$$

and the upper semicontinuity of this multimap follows from Theorems 1.3.11 and 1.3.17. Therefore, condition $(F2)$ is fulfilled.

On the other hand, for each fixed $x \in \mathbb{R}^n$ the map $\psi : I \times \mathbb{R}^m \to \mathbb{R}^n$, $\psi(t, u) = f(t, x, u)$ satisfies the Carathéodory conditions and hence the multimap $F(\cdot, x) = f(\cdot, x, U(\cdot, x)) = \psi(\cdot, U(\cdot, x))$ is measurable and so condition $(F1)$ is also fulfilled.

From Theorems 3.2.6 and 3.2.9 it follows that the set $\Sigma(x_0)$ of all solutions to inclusion (3.26) satisfying initial condition (3.27) (which by the previous theorem is also the set of all trajectories of the control system) is nonempty and compact. Therefore, it contains a trajectory x_* minimizing the functional ϕ. ∎

As one more example, let us consider *time optimal control problem*, i.e., the problem of finding such a control u_* which steers the system from a given initial set $M_0 \subset \mathbb{R}^n$ to a certain *target set* $M \subset \mathbb{R}^n$ in the shortest possible time.

Theorem 3.4.3. *Suppose that for a control system conditions $(f1)$, $(f2)$ and $(U1)$–$(U4)$ are fulfilled; initial set M_0 is compact; the target set M is closed. If there exists a trajectory of the system reaching the set M from the set M_0 at a certain moment t_1, $t_0 < t_1 \leq T$, then this system admits a time optimal control.*

Proof. Denote, as earlier, $\Sigma(x) \subset C(I; \mathbb{R}^n)$ the set of all trajectories of the system emanating from a point $x \in M_0$. From Theorems 3.2.9 and 3.2.12 we know that the multimap Σ is compact-valued and u.s.c. Whence, the set $\Sigma(M_0)$ of all trajectories emanating from M_0 is compact (Theorem 1.2.35).

Consider *the attainability multifunction* $\Pi : I \to K(\mathbb{R}^n)$, $\Pi(t) = \Sigma(M_0)(t)$ and the set $\mathcal{T} = \Pi_-^{-1}(M)$. This set is nonempty since, by assumption, it contains the point t_1. Further, the multifunction Π is u.s.c. (see Example 1.2.22(b)) and hence the set $\mathcal{T} \subset I$ is closed and so contains its lower bound t_* which is just the minimal possible moment of attainability of the set M. ∎

Chapter 4

On some applications

Practical idea can be used from both the right and the left sides.

—*Stanislaw Jerzy Lec*

4.1 Generalized dynamical systems

- If an object moves, it means that it moves in the wrong direction.
- If an object is motionless, it means that it is located in the wrong place.

—*Jerrold's laws of dynamics*

4.1.1 *General properties*

Let X be a certain set. *A dynamical system* on X is a map $q : X \times \mathbb{R} \to X$ satisfying the following conditions:

q1) $q(\cdot, 0) = id_X$, where id_X is the identity map on X;
q2) for each $x \in X$; $t_1, t_2 \in \mathbb{R}$ we have

$$q(q(x, t_1), t_2) = q(x, t_1 + t_2)$$

(the semigroup property).

These axioms are in accordance with the intuitive idea about a stationary law of the motion of a determined system being in a state x at initial moment $t = 0$ and in a uniquely defined state $q(x, \tau)$ at the moment $t = \tau$.

In particular, such dynamical system arises very naturally while the consideration of an autonomous differential equation

$$x'(t) = f(x(t))$$

(where $f : \mathbb{R}^n \to \mathbb{R}^n$) under condition that the Cauchy problem for such equation has a unique solution for each initial value and every such solution is extendable to the whole axis. Indeed, if $x(\cdot)$ is a solution of a given equation satisfying an initial condition $x(0) = x_0$ then the dynamical system q can be defined by means of the translation operator $q : \mathbb{R}^n \times \mathbb{R} \to \mathbb{R}^n$; $q(x_0, t) = x(t)$.

It is easy to see that axioms $(q1)$ and $(q2)$ are fulfilled in this case.

However the application of usual dynamical systems to the study of many processes meets difficulties. Very often it is due to the fact that an initial state of a system does nor define uniquely its further evolution. Precisely such systems arise under consideration of differential equations not satisfying the condition of uniqueness of a solution as well as parametrized differential equations or differential inclusions. This list includes also control systems in the general sense, i.e., systems whose evolution can proceed in different ways depending on the impact of controls.

It is clear that if we wish to describe such objects in terms of dynamical systems then the axiomatic of usual dynamical systems should be revised.

The desired result will be achieved if we admit that the map Q defining a dynamical system is multivalued:

$$Q : X \times \mathbb{R} \to P(X)$$

and obeys the following rules:

$(Q1)$ $Q(\cdot, 0) = id_X$;
$(Q2)$ *for each* $x \in X$ *and* $t_1, t_2 \in \mathbb{R}$, $t_1 \cdot t_2 \geq 0$ *we have:*

$$Q(Q(x, t_1), t_2) = Q(x, t_1 + t_2);$$

$(Q3)$ *for each* $x \in X$, $t \in \mathbb{R}$ *the relation* $y \in Q(x, t)$ *implies* $x \in Q(y, -t)$.

The set $Q(x, \tau)$ which is frequently called *the attainability set* for $\tau \geq 0$ can be naturally interpreted as the set of all states which the system can attain by the moment $t = \tau$ if at the initial moment $t = 0$ it was in the state x. Axiom $(Q3)$ allows to describe the set $Q(x, \tau)$ for $\tau < 0$ as the collection of all states at the moment $t = \tau$ from which the system can attain the state x by the zero moment.

Lemma 4.1.1. *For every* $x \in X$, $t_1, t_2 \in R$ *we have*

$$Q(x, t_1 + t_2) \subset Q(Q(x, t_1), t_2). \tag{4.1}$$

Proof. The case $t_1 \cdot t_2 \geq 0$ is covered by axiom $(Q2)$. Suppose, for example, that $t_1 > 0$, $t_2 < 0$, $t_1 + t_2 > 0$. Then from axiom $(Q2)$ it follows that $Q\left(x, t_1\right) = Q\left(Q\left(x; t_1 + t_2\right), -t_2\right)$.
Then $Q\left(Q\left(x, t_1\right), t_2\right) = Q\left(Q\left(Q\left(x; t_1 + t_2\right), -t_2\right), t_2\right)$.
From axiom $(Q3)$ we get that for every $x \in X$, $t \in \mathbb{R}$:

$$x \in Q\left(Q\left(x, t\right); -t\right).$$

Therefore

$$Q\left(x, t_1 + t_2\right) \subset Q\left(Q\left(Q\left(x, t_1 + t_2\right), -t_2\right), t_2\right),$$

implying inclusion (4.1).

Remaining cases can be considered similarly; we left them to the reader as an exercise. ∎

Lemma 4.1.2. *For every $x \in X$, $t \geq 0$ and $y \in Q\left(x, t\right)$, $0 \leq \tau \leq t$ we have*

$$Q\left(x, \tau\right) \cap Q\left(y, \tau - t\right) \neq \emptyset.$$

Proof. From $y \in Q\left(x, t\right) = Q\left(Q\left(x, \tau\right), t - \tau\right)$ it follows that there exists a point $z \in Q\left(x, \tau\right)$ such that $y \in Q\left(z, t - \tau\right)$.
But then, according to axiom $(Q3)$: $z \in Q\left(y, \tau - t\right)$ that proves the assertion. ∎

To obtain more significant facts concerning generalized dynamical systems we need some additional assumptions concerning the continuity of the multimap Q.

Let X be a Hausdorff topological space. We will suppose that the multimap Q defining the dynamical system has compact values and satisfies, additionally to $(Q1)$–$(Q3)$ the following conditions:

$(Q4)$ *the multimap $Q : X \times \mathbb{R} \to K\left(X\right)$ is upper semicontinuous;*
$(Q5)$ *for every $x \in X$ the multimap $Q\left(x, \cdot\right) : \mathbb{R} \to K\left(X\right)$ is continuous.*

Definition 4.1.3. A multimap $Q : X \times \mathbb{R} \to K\left(X\right)$ satisfying axioms $(Q1) - (Q5)$ defines *a generalized dynamical system (GDS)*.

Generalized dynamical systems are called sometimes *dynamical systems without uniqueness*. To denote a generalized dynamical system, we will use the same symbol Q as for the multimap which defines it.

As an example of a GDS we can consider the system Q_F generated by an autonomous differential inclusion in the space $\mathbb{R}^n$:

$$x'\left(t\right) \in F\left(x\left(t\right)\right). \tag{4.2}$$

Let the multimap $F : \mathbb{R}^n \to Kv(\mathbb{R}^n)$ be u.s.c. and satisfy the sublinear growth condition $\|F(x)\| \le C(1 + \|x\|)$, $x \in \mathbb{R}^n$ for a positive constant C. Then, as we know from Section 3.2, each solution of inclusion (4.2) is extendable onto the whole axis. If to denote, as earlier, by $\Sigma(x)$ the set of all solutions to inclusion (4.2) emanating from a point $x \in \mathbb{R}^n$ at the moment $t = 0$ then the dynamical system $Q_F : \mathbb{R}^n \times \mathbb{R} \to K(\mathbb{R}^n)$ can be defined as

$$Q_F(x,t) = \Sigma(x)(t).$$

The fulfilment of axiom $(Q5)$ for system Q_F follows from Corollary 3.2.10. Property $(Q4)$ needs to be verified.

Lemma 4.1.4. *Dynamical system Q_F satisfies axiom $(Q4)$.*

Proof. From Theorems 3.2.12 and 1.2.35 it follows that the multimap Q_F is locally compact and hence, by virtue of Theorem 1.2.32 it is sufficient to verify its closedness. Let $\{x_n\} \subset \mathbb{R}^n$, $x_n \to x_0$, $\{t_n\} \subset \mathbb{R}$, $t_n \to t_0$, $y_n \in Q_F(x_n, t_n)$, $y_n \to y_0$.

Take an interval $[0, T] \subset \mathbb{R}$ containing the sequence $\{t_n\}$ and the limit point t_0. Then by definition there exists a sequence of solutions $\{\tilde{x}_n\} \subset C([0, T]; \mathbb{R}^n)$ to inclusion (4.2) such that $\tilde{x}_n(0) = x_n$ and $\tilde{x}_n(t_n) = y_n$ for all $n = 1, 2, \ldots$ The sequence of solutions $\{\tilde{x}_n\}$ is relatively compact, so, without loss of generality, we can assume that it converges to a solution $\tilde{x}_0 \subset C([0, T]; \mathbb{R}^n)$ for which we obviously have $\tilde{x}_0(0) = x_0$, $\tilde{x}_0(t_0) = y_0$. Whence $y_0 \in Q_F(x_0, t_0)$. ∎

Consider some properties of generalized dynamical systems. We will need the following auxiliary statement.

Lemma 4.1.5. *Let X, Y, Z be topological spaces; a multimap $F : X \times Y \to P(Z)$ be u.s.c. Suppose that for every $x \in X$ the multimap $F(x, \cdot) : Y \to P(Z)$ is continuous. Then for each $A \in K(X)$ the multimap $F_A : Y \to P(Z)$,*

$$F_A(y) = F(A, y)$$

is continuous.

Proof. The upper semicontinuity of the multimap F_A follows from the fact that it can be represented as the composition of the continuous multimap $G : Y \to K(X \times Y)$, $G(y) = A \times \{y\}$ and the multimap F (see Theorem 1.3.11).

On the other hand, if $V \subset Z$ is an open set then $(F_A)_-^{-1}(V)$ is also open since $(F_A)_-^{-1}(V) = \bigcup_{a \in A} [F(a, \cdot)]_-^{-1}(V)$.

Therefore, the multimap F_A is lower semicontinuous. ∎

Theorem 4.1.6. *Let Q be a GDS defined by a multimap $Q : X \times \mathbb{R} \to K(X)$; $A \subset X$ a connected compact set. Then for each $t_1, t_2 \in \mathbb{R}$, $t_1 \leq t_2$, $0 \in [t_1, t_2]$ the set $Q(A, [t_1, t_2])$ is connected and compact.*

Proof. The compactness of the set $Q(A, [t_1, t_2])$ follows from axiom $(Q4)$ and Theorem 1.2.35. By virtue of axioms $(Q4)$, $(Q5)$ and Lemma 4.1.5 the multimap $Q_A : \mathbb{R} \to K(X)$, $Q_A(t) = Q(A, t)$ is continuous. The assertion now follows from the fact that the set $Q_A(0) = Q(A, 0) = A$ is connected by assumption and Theorem 1.2.37(ii). ∎

Definition 4.1.7. *A trajectory of a generalized dynamical system Q on an interval $[a, b] \subset \mathbb{R}$ is a function $\varphi : [a, b] \to X$ satisfying for every $t_0, t_1 \in [a, b]$ the relation*

$$\varphi(t_2) \in Q(\varphi(t_1), t_2 - t_1). \tag{4.3}$$

Let us indicate some properties of trajectories of generalized dynamical systems.

The next statement follows directly from Definition 4.1.7 and axioms $(Q1)$, $(Q5)$.

Lemma 4.1.8. *Each trajectory of a GDS is a continuous function.*

The following statement on the join of trajectories will be useful in the sequel.

Lemma 4.1.9. *if φ_1 is a trajectory of a GDS Q on the interval $[a, b]$ and φ_2 is its trajectory on $[b, c]$ such that $\varphi_1(b) = \varphi_2(b)$ then the function $\varphi : [a, c] \to X$,*

$$\varphi(t) = \begin{cases} \varphi_1(t), & a \leq t \leq b; \\ \varphi_2(t), & b < t \leq c, \end{cases}$$

is the trajectory of the system Q on the interval $[a, c]$.

Proof. Take $t_1, t_2 \in [a, c]$. If $t_1, t_2 \in [a, b]$ or $[b, c]$ then the fulfilment of relation (4.3) for the function φ is obvious. Let, for example, $t_1 \in [a, b]$, $t_2 \in [b, c]$. Then $\varphi(t_2) = \varphi_2(t_2) \in Q(\varphi_2(b), t_2 - b) = Q(\varphi_1(b), t_2 - b) \subset Q(Q(\varphi_1(t_1), b - t_1), t_2 - b) = Q(\varphi_1(t_1), t_2 - t_1) = Q(\varphi(t_1), t_2 - t_1)$.

The case $t_1 \in [b, c]$, $t_2 \in [a, b]$ can be considered analogously. ∎

Theorem 4.1.10. *Let Q be a GDS. For each $a \le b$, $x_1 \in Q(x_0, b - a)$ there exists a trajectory φ on the interval $[a, b]$ such that $\varphi(a) = x_0$, $\varphi(b) = x_1$.*

Proof. For simplicity, take $a = 0$, $b = 1$. At first, let us define the values of the trajectory φ on the countable dense subset $J \subset [0, 1]$: $J = \left\{0, 1, \frac{1}{2}, \frac{1}{4}, \frac{3}{4}, \frac{1}{8}, \frac{3}{8}, ..., \frac{p}{2^q}, ...\right\}$ by means of the following inductive process.

Set $\varphi(0) = x_0$, $\varphi(1) = x_1$ and select $\varphi\left(\frac{1}{2}\right)$ arbitrarily in the set $Q\left(\varphi(0), \frac{1}{2}\right) \cap Q\left(\varphi(1), -\frac{1}{2}\right)$ which is nonempty by virtue of Lemma 4.1.2. Further, if $\varphi\left(\frac{p}{2^q}\right)$ and $\varphi\left(\frac{p+1}{2^q}\right)$ are constructed already then we select $\varphi\left(\frac{2p+1}{2^{q+1}}\right)$ in such a way that

$$\varphi\left(\frac{2p+1}{2^{q+1}}\right) \in Q\left(\varphi\left(\frac{p}{2^q}\right), \frac{1}{2^{q+1}}\right) \cap Q\left(\varphi\left(\frac{p+1}{2^q}\right), -\frac{1}{2^{q+1}}\right).$$

The nonemptyness of the intersection also follows from Lemma 4.1.2. So, we defined $\varphi(t)$ for all $t \in J$ and, moreover, from Lemma 4.1.1 it follows that $\varphi(t'') \in Q(\varphi(t'), t'' - t')$ for all $t', t'' \in J$.

If $t \in [0, 1] \setminus J$ then consider the set

$$K(t) = \bigcap_{t', t'' \in J; t' < t < t''} Q(\varphi(t'), t - t') \cap Q(\varphi(t''), t - t'').$$

From Lemmas 4.1.2 and 4.1.1 it follows that

$$\{Q(\varphi(t'), t - t') \cap Q(\varphi(t''), t - t'')\}_{t', t'' \in J, t' < t < t''}$$

is a family of sets with the finite intersection property and hence the set $K(t)$ is nonempty. Let us show that it is a singleton.

Indeed, if $y \in K(t)$ then for all $t'' \in J$, $t < t''$ we have $\varphi(t'') \in Q(y, t'' - t)$ from where it follows that for an arbitrary neighborhood $U(y) \subset X$ of a point y there exists $\delta > 0$ such that $\varphi(t'') \in U(y)$ for all $t'' \in J$, $0 < t'' - t < \delta$.

Now set $\varphi(t) = K(t)$ for all $t \in [0, 1] \setminus J$.

So defined function φ is the desired trajectory. Indeed, if $t_1, t_2 \in J$ then, by construction $\varphi(t_2) \in Q(\varphi(t_1), t_2 - t_1)$.

If, for example, $t_1, t_2 \in [0, 1] \setminus J$, $t_1 < t_2$ then take $t' \in J$, $t_1 < t' < t_2$. We have

$$\varphi(t_1) = K(t_1) \in Q(\varphi(t'), t_1 - t'),$$

hence

$$\varphi(t') \in Q(\varphi(t_1), t' - t_1).$$

But then

$$\varphi\left(t_2\right) = K\left(t_2\right) \in Q\left(\varphi\left(t'\right), t_2 - t_1\right) \subset Q\left(Q\left(\varphi\left(t_1\right), t' - t_1\right), t_2 - t'\right) =$$

$$= Q\left(\varphi\left(t_1\right), t_2 - t_1\right).$$

The remaining cases can be considered analogously. ∎

Applying Lemma 4.1.9 we get the following statement.

Corollary 4.1.11. *Let Q be a GDS and sequences $\{t_i\}_{i=0}^{n} \subset \mathbb{R}$, $t_0 < t_1 <$ ... $< t_n$; $\{x_i\}_{i=0}^{n} \subset X$ satisfy condition*

$$x_k \in Q\left(x_{k-1}, t_k - t_{k-1}\right), \quad 1 \leq k \leq n.$$

Then there exists a trajectory φ on the interval $[t_0, t_n]$ such that $\varphi\left(t_i\right) = x_i$ $(i = 0, 1, ..., n)$.

Let Q be a GDS given by a multimap $Q : X \times \mathbb{R} \to K\left(X\right)$, $A \subset X$, $[a, b] \subset \mathbb{R}$. Denote by $S\left(Q, A, [a, b]\right)$ the set of all trajectories φ of Q on the interval $[a, b]$ such that $\varphi\left(a\right) \in A$.

Theorem 4.1.12. *Let (X, ϱ) be a metric space, Q a GDS given by a multimap $Q : X \times \mathbb{R} \to K\left(X\right)$. Then for every $A \in K\left(X\right)$, $[a, b] \subset \mathbb{R}$ the set $S\left(Q, A, [a, b]\right)$ is compact with respect to the uniform convergence topology.*

Proof. First of all, notice that the total range for the family of trajectories $S\left(Q, A, [a, b]\right)$, as it is easy to see, coincides with the compact set $Q\left(A, [0, b - a]\right) \subset X$.

This gives the opportunity to use the Arzela–Ascoli theorem (see Chapter 0) to prove the relative compactness of the set $S\left(Q, A, [a, b]\right)$ in the space $C\left([a, b]; X\right)$.

Let us show that the set of functions $S\left(Q, A, [a, b]\right)$ is equicontinuous. Take any $\varepsilon > 0$. By virtue of axioms $(Q1)$ and $(Q4)$ for each point $x \in Q\left(A, [0, b - a]\right)$ there exist its neighborhood $U\left(x\right)$ and a number $\delta_x > 0$ such that

$$Q\left(U\left(x\right), \left[-\frac{\delta_x}{2}, \frac{\delta_x}{2}\right]\right) \subset B_{\frac{\varepsilon}{2}}\left(x\right).$$

Notice that from the construction it follows that $U\left(x\right) \subset B_{\frac{\varepsilon}{2}}\left(x\right)$.

The system of neighborhoods $\{U\left(x\right)\}_{x \in Q\left(A, [0, b-a]\right)}$ forms a cover of the set $Q\left(A, [0, b - a]\right)$. Select a finite subcover $\{U\left(x_i\right)\}_{i=1}^{n}$ and denote $\delta = \min_{1 \leq i \leq n} \delta_{x_i}$.

Now, let φ be any function from the family $S(Q, A, [a, b])$; $t_1, t_2 \in [a, b]$, $|t_2 - t_1| < \delta$. Let $\varphi(t_1) \in U(x_k)$, $1 \le k \le n$.

Then

$$\varphi(t_2) \in Q(\varphi(t_1), t_2 - t_1) \subset B_{\frac{\varepsilon}{2}}(x_k).$$

Since also $\varphi(t_1) \in B_{\frac{\varepsilon}{2}}(x_k)$, we have $\varrho(\varphi(t_1), \varphi(t_2)) < \varepsilon$ giving the equicontinuity of $S(Q, A, [a, b])$.

To conclude the proof we should demonstrate that the set $S(Q, A[a, b])$ is closed.

Let

$$\{\varphi_j\}_{i=1}^{\infty} \subset S(Q, A, [a, b]), \quad \|\varphi_j - \varphi_0\|_C \to 0.$$

If $t_1, t_2 \in [a, b]$ then for each $j \in \mathbb{N}$ we have

$$\varphi_j(t_2) \in Q(\varphi_j(t_1), t_2 - t_1).$$

Since the multimap Q is closed (Theorem 1.2.29) we can pass in this inclusion to the limit as $j \to \infty$ and to obtain

$$\varphi_0(t_2) \in Q(\varphi_0(t_1), t_2 - t_1).$$

Therefore, the function φ_0 is also the trajectory of the system Q. ∎

Let us return to the consideration of a GDS generated by autonomous differential inclusion (4.2) under given above conditions on its right-hand part. From the definition it is clear that every solution of this inclusion on an interval $[a, b] \subset \mathbb{R}$ is a trajectory of the corresponding GDS. So, it seems very natural to ask if, conversely, every trajectory of such a system is a solution of the corresponding differential inclusion.

Theorem 4.1.13. *Each trajectory of the GDS Q_F generated by differential inclusion (4.2) on an interval $[a, b]$ is a solution of this inclusion on the same interval.*

Proof. For simplicity, set $a = 0, b = 1$. Let $\varphi : [0, 1] \to \mathbb{R}^n$ be a trajectory of Q_F. Construct a sequence $\{\psi_k\}_{k=0}^{\infty}$; $\psi_k : [0, 1] \to \mathbb{R}^n$ of solutions of differential inclusion (4.2) in such a way that the function ψ_k coincides with the function φ at points of the form $\left\{ \frac{p}{2^k} : p = 0, 1, ..., 2^k \right\}$. Such construction is feasible since the inclusion

$$\varphi\left(\frac{p+1}{2^k}\right) \in Q_F\left(\varphi\left(\frac{p}{2^k}\right), \frac{1}{2^k}\right), \quad p = 0, 1, ..., 2^k - 1$$

implies the existence of a solution ψ_k^p of (4.2) on the interval $\left[\frac{p}{2^k}, \frac{p+1}{2^k}\right]$ such that

$$\psi_k^p\left(\frac{p}{2^k}\right) = \varphi\left(\frac{p}{2^k}\right), \quad \psi_k^p\left(\frac{p+1}{2^k}\right) = \varphi\left(\frac{p+1}{2^k}\right).$$

It is clear that the solution $\psi_k : [0,1] \to \mathbb{R}^n$,

$$\psi_k(t) = \psi_k^p(t), \quad t \in \left[\frac{p}{2^k}, \frac{p+1}{2^k}\right]$$

satisfies the desirable condition.

Since $\psi_k(0) = \varphi(0)$ for all k, the family of solutions $\{\psi_k\}$ is compact in the space $C\left([0,1]\,;\mathbb{R}^n\right)$ (Theorem 3.2.9). Hence we can extract from the sequence $\{\psi_l\}_{k=0}^{\infty}$ a subsequence uniformly converging to a solution ψ. From the construction of the sequence $\{\psi_k\}$ it follows that the functions ψ and φ coincide on a countable dense subset

$$J = \left\{\frac{p}{2^k} : 0 \leq k < \infty;\ 0 \leq p \leq 2^k\right\}$$

of the interval $[0,1]$. But, by virtue of the continuity of both functions it means that $\varphi = \psi$ on the whole interval $[0,1]$. ∎

4.1.2 Rest points of one-sided dynamical systems

In this section we will consider the problem on the existence of a constant trajectory or *a rest point* for a generalized dynamical system. Since as a main tool we will use fixed point theorems considered in the second chapter, we will assume for simplicity that a multimap defining a dynamical system is convex-valued. The application of other methods, in particular, asymptotic fixed point theorems for multimaps with non-convex values opens the opportunity to study rest points or attractors for more general classes of dynamical systems (see, for example, [314], [157], [235], [325] and other works).

We slightly change the above described axiomatics in conformity to the problem under investigation.

Let X be a metric space; $\mathbb{R}_+ = [0, \infty)$.

Definition 4.1.14. A multimap $G : X \times \mathbb{R}_+ \to C(X)$ defines *an one-sided generalized dynamical system (OGDS)*, if it satisfies the following conditions:

(G1) $G(\cdot, 0) = id_X$;

(G2) *for each* $x \in X$, $t_1, t_2 \in \mathbb{R}_+$:
$$G\left(G\left(x, t_1\right), t_2\right) \subset G\left(x, t_1 + t_2\right);$$
(G3) G *is u.s.c.*

Definition 4.1.15. A point $x_\star \in X$ is called *a rest point* of an OGDS G if $x_\star \in G\left(x_\star, t\right)$ for all $t \in \mathbb{R}_+$.

For a fixed $t \in \mathbb{R}_+$ define the multimap $G_t : X \to C\left(X\right)$ by the equality $G_t\left(x\right) = G\left(x, t\right)$. We use the following sufficient condition for the existence of a rest point (see [314]).

Lemma 4.1.16. *Let a sequence* $\{t_k\}$, $t_k > 0$, $t_k \to 0$ *is such that each multimap* G_{t_k} *has a fixed point* x_k. *If* $x_\star$ *is a limit point of the sequence* $\{x_k\}$ *then* $x_\star$ *is a rest point of* G.

Proof. By using axiom (G2) we have for every fixed $t > 0$ and $k \in \mathbb{N}$:
$$G\left(x_k, t\right) \supset G\left(G\left(x_k, t_k\right), t - t_k\right) \supset G\left(x_k, t - t_k\right) \supset G\left(G\left(x_k, t_k\right), t - 2t_k\right) \supset \ldots$$
$$\ldots \supset G\left(x_k, t - \left[\frac{t}{t_k}\right] t_k\right).$$
Assume that $x_\star \notin G\left(x_\star, t\right)$. Then the sets $\{x_\star\} = G\left(x_\star, 0\right)$ and $G\left(x_\star, t\right)$ can be separated with disjoint open neighborhoods V_0 and V_1. By axiom (G3) for sufficiently large k we will have
$$G\left(x_k, t - \left[\frac{t}{t_k}\right] t_k\right) \subset V_0, \quad G\left(x_k, t\right) \subset V_1,$$
that is impossible. ∎

We can present now the following rest point theorems.

Theorem 4.1.17. *Let* M *be a convex compact subset of a Banach space; a multimap* $G : M \times \mathbb{R}_+ \to Kv\left(M\right)$ *defines an OGDS. Then* G *has a rest point.*

Proof. By virtue of Theorem 2.2.22, each multimap $G_t : M \to Kv\left(M\right)$, $t > 0$ has a fixed point. Choosing an arbitrary sequence $\{t_k\}$, $t_k > 0$, $t_k \to 0$, a corresponding sequence of fixed points x_k of the multimaps G_{t_k} and a limit point $x_\star$ of this sequence, we can apply Lemma 4.1.16. ∎

Theorem 4.1.18. *Let* $U \subset \mathbb{R}^n$ *be a bounded open set. Suppose that a multimap* $G : \overline{U} \times \mathbb{R}_+ \to Kv\left(\overline{U}\right)$ *defines an OGDS and for some* $T > 0$ *the set* $G\left(\overline{U}, T\right)$ *is contained in a convex open set* U_0, $\overline{U}_0 \subset U$. *Then the OGDS* G *has a rest point* $x_\star \in U_0$.

Proof. For a sequence $t_k = \frac{T}{k}$ consider the multimaps $G_{t_k} : \overline{U} \to Kv\left(\overline{U}\right)$. By virtue of axiom $(G2)$, for each $k \in \mathbb{N}$ we have

$$G_{t_k}^k\left(\overline{U}\right) \subset G\left(\overline{U}, T\right) \subset U_0.$$

Applying Theorem 2.2.28 to the multimap G_{t_k} we obtain that it should have a fixed point $x_k \in U_0$. Take again a limit point x_* of the sequence $\{x_k\}$ and apply Lemma 4.1.16. ∎

4.2 On applications in theory of games and mathematical economics

4.2.1 *Optimal strategies in zero-sum games*

> If a man says: "It is a silly childish game" it means that it is a game in which he loses to his wife.
>
> —*Epperson's law*

We are already familiar (see Example 1.1.15) with the simplest model of the two-person game. So we suppose that for each of the two players there is given his space of strategies. We denote these sets X and Y respectively. The model also includes the game rules, multimaps $A : Y \to P(X)$ and $B : X \to P(Y)$ characterizing best responses of each of the players to a strategy applied by the opponent.

What should each player now be guided by when choosing strategies? Are there any strategies which should be called optimal?

Definition 4.2.1. A pair of strategies $x_0 \in X$, $y_0 \in Y$ forms *the equilibrium situation* if the following inclusions hold true:

$$\begin{cases} x_0 \in A\left(y_0\right) \\ y_0 \in B\left(x_0\right). \end{cases}$$

This game position really can be characterized as equilibrium: the strategy x_0 is one of the strongest responses to the strategy y_0 applied by the second player whereas y_0 is also among the strongest objections to the strategy x_0. In this sense these strategies are optimal and, if the players have no additional information about the intentions of the opponent, they should stick to equilibrium strategies.

In the case of a zero-sum game given with the help of a game function $f : X \times Y \to \mathbb{R}$ (see Example 1.1.15), the first player applying the optimal strategy x_0 guarantees his payoff not less than $f(x_0, y_0)$ under any response of the opponent, and at the same time the second player applying the strategy y_0, will not give the first one an opportunity to increase his payoff above this value. In this case the number $f(x_0, y_0)$ is called *the value of the game.*

When is it possible to guarantee the existence of optimal strategies? Let us pay attention to the fact that a pair of strategies x_0, y_0 forms the equilibrium situation if and only if (x_0, y_0) is a fixed point of the multimap $A \times B : X \times Y \longrightarrow P(X \times Y)$ defined as $(A \times B)(x, y) = A(y) \times B(x)$. This gives the opportunity to apply the Bohnenblust–Karlin fixed point theorem (Theorem 2.2.22) to prove the existence of optimal strategies. Indeed, the following statement holds true.

Theorem 4.2.2. *Let the spaces of strategies X and Y be convex compact subsets of Banach spaces E_0 and E_1 respectively. If the game rules A and B are u.s.c. multimaps with convex closed values then the game has an equilibrium situation.*

Proof. Applying Theorem 1.3.17, it is easy to see that the multimap $A \times B : X \times Y \to Kv(X \times Y)$ of the convex compact subset $X \times Y$ of the Banach space $E_0 \times E_1$ into itself is u.s.c. and hence satisfies conditions of Theorem 2.2.22. ∎

To make the above abstract result more concrete, consider the following theorem on optimal strategies in a zero-sum game which is its particular case.

Theorem 4.2.3. *Let the spaces of strategies X and Y in a zero-sum game of two persons be convex compact subsets of Banach spaces E_0 and E_1 respectively and a payoff function $f(x, y)$ defined on $X \times Y$ be continuous, convex in y for each $x \in X$ and concave in x for each $y \in Y$, i.e.,*

$$f(x, (1 - \lambda) y' + \lambda y'') \leq (1 - \lambda) f(x, y') + \lambda f(x, y'')$$

for each $x \in X$; $y', y'' \in Y$; $\lambda \in [0, 1]$ and

$$f((1 - \lambda) x' + \lambda x'', y) \geq (1 - \lambda) f(x', y) + \lambda f(x'', y)$$

for each $x', x'' \in X$; $y \in Y$; $\lambda \in [0, 1]$. Then the game has an equilibrium situation.

Proof. From the theorem of maximum (Theorem 1.3.29) it follows that the game rules A and B in this case are compact-valued u.s.c. multimaps and the convexity of their values follows from the imposed conditions (**verify it!**). ∎

At first glance, the sphere of practical applications of the above theorems seems to be very narrow first of all due to the extremely restrictive condition of convexity of the spaces of strategies. Indeed, in real situations the number of admissible strategies for each player although can be very large, in fact it is finite. Even in chess the number of all possible positions is not unlimited!

John von Neumann suggested the following interpretation which overcomes this difficulty. Suppose that a player has a finite number of strategies $x_1, ..., x_k$ which are called *pure strategies*. Consider the set $X \subset \mathbb{R}^k$

$$X = \left\{ (\lambda_1, ..., \lambda_k) \mid \lambda_i \geq 0, \ i = 1, ..., k; \ \sum_{i=1}^{k} \lambda_i = 1 \right\}$$

which is called *the standard simplex* of the space $\mathbb{R}^k$. The elements of this (obviously convex and compact) set can be interpreted as *mixed strategies*, i.e., we suppose that a player chooses not a single pure strategy x_i but uses, generally speaking, all strategies and determines only the probabilities λ_i, $i = 1, ..., k$, in accordance with which he will use them in a number of rounds of the game. In this case the obtained payoff is also considered in an averaged form.

As an example, let us consider a widespread case of *the matrix game*, i.e., a two-person zero-sum game in which each of the players has only a finite number of pure strategies. Let the strategies $x_1, ..., x_n$, be available for the first player whereas his opponent has at his disposal the strategies $y_1, ..., y_m$. In this case it is convenient to present the payoff function in the form of *the game matrix*

$$\begin{pmatrix} a_{11} & a_{12} & \cdots & a_{1m} \\ a_{21} & a_{22} & \cdots & a_{2m} \\ . & . & \cdots & . \\ a_{n1} & a_{n2} & \cdots & a_{nm} \end{pmatrix}$$

An element a_{ij} of the matrix characterizes the payoff of the first player if he applies a strategy x_i $(i = 1, ..., n)$ on which the second player responses by a strategy y_j $(j = 1, ..., m)$.

Now, if the first player is applying a mixed strategy of the form $\tilde{x} = (\lambda_1, ..., \lambda_n)$ and the second player uses a mixed strategy $\tilde{y} = (\mu_1, ..., \mu_m)$ then the payoff of the first player is the random variable which takes the values a_{ij} with the probabilities $\lambda_i \mu_j$. Then the payoff of the first player if these strategies are applied is nothing but the mathematical expectation of this variable, i.e.,

$$f(\widetilde{x}, \widetilde{y}) = \sum_{i=1}^{n} \sum_{j=1}^{m} a_{ij} \lambda_i \mu_j.$$

It is clear that this bilinear form satisfies the conditions of Theorem 4.2.3 and so we obtain the following statement (*the von Neumann–Nash equilibrium theorem*)

Theorem 4.2.4. *Each matrix game has an equilibrium situation in mixed strategies.*

4.2.2   *An equilibrium in a model of a competitive economics*

> An economic problem: how to take away from everybody in order to add to everyone.
>
> *—Henryk Jagodziński*

The questions of the functioning of decentralized economical systems have been of great scientific interest for a long time. In these systems, each of the participants acts exclusively in his own interests having no any knowledge about the global state of the economics or the information about the behavior of other contestants. How can such economics act? This paradoxical property was described for the first time by a famous Scottish economist Adam Smith who mentioned in his book "The Wealth of Nations" (1776):

"As every individual, therefore, endeavours as much as he can both to employ his capital in the support of domestic industry, and so to direct that industry that its produce may be of the greatest value, every individual necessarily labours to render the annual revenue of the society as great as he can. He generally, indeed, neither intends to promote the public interest, nor knows how much he is promoting it. By preferring the support of domestic to that of foreign industry, he intends only his own security; and by directing that industry in such a manner as its produce may be of the greatest value, he intends only his own gain, and he is in this, as in many other cases, led by an invisible hand to promote an end which was no part of his intention. Nor is it always the worse for the society that it was not part of it. By pursuing his own interest he frequently promotes that of the society more effectually than when he really intends to promote it."

But Smith did not explain how this famous invisible hand rules, and moreover, he did not give any rigorous arguments in favor of its existence.

A French economist Leon Walras tried to do this a hundred years later, suggesting that the price system can actually be such a controller. The main idea of Walras was that under a certain price system individual plans become compatible with each other. Such economical situation is called *the competitive equilibrium.*

However, the Walras hypothesis remained for many years only as a brilliant conjecture and it was strictly justified only in the fifties of the last century, when the corresponding mathematical apparatus was designed. Its essential part forms the theory of multivalued maps.

Consider the mathematical aspect of the problem dwelling on a model close to those suggested by K.J. Arrow, G. Debreu and L.W. McKenzie (see, for example, [236], [273], [299], [319]).

Let an economical system contain n various types of goods. The system includes k customers for which of them a convex compact set of available goods $X^i \subset R^n$ $(i = 1, ...k)$ *(the customer set)* is given. The choice of each customer is determined by its *utility index* $u^i : R^n \to \mathbb{R}$ $(i = 1, ..., k)$. Each function u^i is supposed to be continuous and satisfying the condition:

the set
$$\left\{ x^i \in X^i \mid u^i\left(x^i\right) \geq u^i\left(\overline{x}^i\right) \right\}$$
is convex for all $\overline{x}^i \in X^i$.

The last demand has a clear economical sense: if certain customer's collections are more useful than a given one then their convex combination has the same property.

Let the collection $(p_1, ..., p_n)$ characterizes the prices on available goods. Since we are interested only in relations between prices but not their absolute values, we will suppose that the set of all prices fills *the price simplex* $\Delta \subset R^n : (p_1, ..., p_n), p_i \geq 0, \sum_{i=1}^n p_i = 1$.

If the collection of goods $x \in R^n$ is given then the cost of this collection under given prices $p \in \Delta$ is characterized by the inner product $\langle p, x \rangle$.

The production potentialities of the system are characterized by the presence of m enterprises-producers. It is supposed that each of them can produce any collection of goods from *the technological set* $Y^j \subset R^n$ $(j = 1, ..., m)$ which is assumed to be convex and compact. Trying to maximize his profit under the given prices, in fact the j-th producer will make his selection in the set
$$\Psi^j\left(p\right) = \left\{ y \in Y^j \mid \langle p, y \rangle = \max_{y' \in Y^j} \langle p, y' \rangle = \pi_j\left(p\right) \right\}.$$
We will assume that each set Y^j contains zero that corresponds to the inactivity and then all marginal functions π_j are nonnegative. So, we get

for each $j = 1, ..., m$ the multimap $\Psi^j : \Delta \to Kv(R^n)$, which is called *productive multifunction of j-th producer*. From the theorem of maximum (Theorem 1.3.29) it follows that all productive multifunctions Ψ^j are u.s.c. and all marginal functions π_j are continuous.

The choice of each consumer is affected by financial constraints. For their description, let us denote by the symbol θ_{ij} the part of the profit of the j-th producer transferred to the i-th consumer (we can talk, for example, about shareholders). It is supposed that $\sum_{i=1}^{k} \theta_{ij} = 1$ (such "altruism" can be completely explained if we consider producers as consumers as well).

To include merchants in this model, let us suppose that the i-th consumer (i = 1,...,k) possesses a collection of goods $\omega^i \in R^n$ for the trade. So, under given prices $p \in \Delta$, the means of i-th consumer composed of trade income and producers contributions may be estimated as

$$\alpha_i(p) = \langle p, \omega^i \rangle + \sum_{j=1}^{m} \theta_{ij}\pi_j(p).$$

It is clear that all the functions α_i are continuous.

The main assumption which is imposed in the described model may be characterized as *the Walras law* which provides to each of the consumers the opportunity "to live within his means and even to save up something". Speaking more precisely, it is supposed that for all prices $p \in \Delta$ every set X^i contains at least one element $x^i = x^i(p)$ such that

$$\langle p, x^i \rangle < \alpha_i(p).$$

To realize this condition it is sufficient to suppose the existence in each set X^i of an element x^i such that $x^i < \omega^i$ in the sense of the component-wise comparison of vectors.

Therefore, the budget restrictions separate for each consumer the set of goods which are really available to him:

$$W^i(p) = \left\{ x \in X^i \mid \langle p, x \rangle \le \alpha_i(p) \right\}.$$

It is clear that by virtue of the above assumptions the set $W^i(p)$ ($i = 1, ..., k$) is nonempty, convex and compact for all $p \in \Delta$.

Notice the following important fact.

Lemma 4.2.5. *Each multimap $W^i : \Delta \to Kv(\mathbb{R}^n)$, ($i = 1, ..., k$) is continuous.*

Proof. Each multimap W^i can be represented as the intersection $W^i = F_0^i \cap F_1^i$ of the constant multimap $F_0^i : \Delta \to Kv(\mathbb{R}^n)$,

$$F_0^i(p) \equiv X^i$$

and the multimap $F_1^i : \Delta \to Cv(\mathbb{R}^n)$,

$$F_1^i(p) = \{x \in \mathbb{R}^n \mid \langle p, x \rangle \le \alpha_i(p)\}.$$

It is easy to see that the multimap F_1^i is closed. Since the multimap F_0^i is continuous, by virtue of Theorem 1.3.3, the multimap W^i is u.s.c.

Now, it is also not difficult to verify **(do it!)** that F_1^i is quasi-open and hence, by Theorem 1.3.8 it is l.s.c. Further, by the Walras law at each point $p \in \Delta$ we have

$$F_0^i(p) \cap int F_1^i(p) \ne \emptyset.$$

So, by applying Theorem 1.3.10 we come to the conclusion that W^i is u.s.c. and hence continuous. ∎

Notice now that each i-th consumer will select in the set $W^i(p)$ of available goods the most useful ones:

$$\Phi^i(p) = \left\{x \in W^i(p) \mid u^i(x) = \max_{x' \in W^i(p)} u^i(x')\right\}.$$

Due to conditions imposed on the utility functions, each $\Phi^i(p)$; $i = 1, ..., k$; $p \in \Delta$ is a nonempty convex compact set. By using Lemma 4.2.5 and applying the theorem of maximum (Theorem 1.3.29) we deduce that the multimaps $\Phi^i : \Delta \to Kv(\mathbb{R}^n)$ $(i = 1, ..., k)$ are u.s.c. The multimap Φ^i is called *the demand multifunction* of the i-th consumer.

Consider now:

the total productive multifunction

$$\Psi(p) = \sum_{j=1}^{m} \Psi^j(p);$$

the total demand multifunction

$$\Phi(p) = \sum_{i=1}^{k} \Phi^i(p);$$

and *the total trade vector*

$$\omega = \sum_{i=1}^{k} \omega^i.$$

Definition 4.2.6. *The* competitive equilibrium *in the model under consideration is the triple*

$$(p_*, x_*, y_*), \ p_* \in \Delta, \ x_* \in \Phi(p_*), \ y_* \in \Psi(p_*),$$

satisfying the relation

$$y_* + \omega \geq x_*.$$

We see that the equilibrium in the given model is achieved under such prices for which all the demand in the system is completely satisfied and besides that each of participants is optimizing only his own interests, not coordinating them with other partners or competitors.

To prove the existence of a competitive equilibrium in the given model we need the following abstract result which is called *the Gale–Nikaido–Debreu theorem on excess demand.*

Theorem 4.2.7. *Let a u.s.c. multimap $T : \Delta \to Kv(\mathbb{R}^n)$ satisfy the condition*

$$\langle p, z \rangle \leq 0$$

for all $p \in \Delta$ and $z \in T(p)$. Then there exists $p^* \in \Delta$ and $z^* \in T(p^*)$ such that $z^* \leq 0$.

Proof. Applying Theorem 1.2.35 we can verify that the set $Z = \overline{co}T(\Delta)$ is compact. Consider now the multimap $S : Z \to Kv(\Delta)$,

$$S(z) = \left\{ p \in \Delta \mid \langle p, z \rangle = \max_{p' \in \Delta} \langle p', z \rangle \right\}.$$

By virtue of the theorem of maximum the multimap S is u.s.c.

Define now the multimap $F : \Delta \times Z \to Kv(\Delta \times Z)$ in the following way:

$$F(p, z) = S(z) \times T(p).$$

From Theorem 1.3.17 it follows that the multimap F is u.s.c. and so we can apply to it the Kakutani fixed point theorem (Theorem 2.2.22) and get the existence of $p^* \in \Delta$ and $z^* \in Z$ such that $p^* \in S(z^*)$, $z^* \in T(p^*)$.

But then, on the one hand, we have $\langle p^*, z^* \rangle \leq 0$ by the condition of the theorem and, on the other hand, $\langle p, z^* \rangle \leq \langle p^*, z^* \rangle$ for each $p \in \Delta$ by the definition of the multimap S. This means that $\langle p, z^* \rangle \leq 0$ for each $p \in \Delta$ that implies $z^* \leq 0$. ∎

Now we are in position to prove the main statement of this section.

Theorem 4.2.8. *Under the above assumptions the given model has a competitive equilibrium.*

Proof. From Theorem 1.3.20 it follows that *the excess demand multifunction* $T : \Delta \to Kv(\mathbb{R}^n)$ defined as

$$T(p) = \Phi(p) - \Psi(p) - \omega,$$

is u.s.c. Let us show that it satisfies the conditions of Theorem 4.2.7.

In fact, take arbitrary $p \in \Delta$ and $z \in T(p)$. Then we have

$$z = x - y - \omega,$$

where $x = \sum_{i=1}^{k} x^i$, $x^i \in \Phi^i(p)$, $i = 1, ..., k$; and $y = \sum_{j=1}^{m} y^j$, $y^j \in \Psi^j(p)$, $j = 1, ..., m$.

By virtue of optimality of the vectors y^j $(j = 1, ..., m)$ we get, for each $i = 1, ..., k$, the relations

$$\langle p, x^i \rangle \leq \langle p, \omega^i \rangle + \sum_{j=1}^{m} \theta_{ij} \langle p, y^j \rangle.$$

Summing these inequalities in i we get

$$\langle p, x \rangle \leq \langle p, \omega \rangle + \sum_{i=1}^{k} \sum_{j=1}^{m} \theta_{ij} \langle p, y^j \rangle$$

$$= \langle p, \omega \rangle + \sum_{j=1}^{m} \sum_{i=1}^{k} \theta_{ij} \langle p, y^j \rangle = \langle p, \omega \rangle + \sum_{j=1}^{m} \langle p, y^j \rangle$$

$$= \langle p, \omega \rangle + \langle p, y \rangle = \langle p, \omega + y \rangle,$$

that yields $\langle p, z \rangle \leq 0$. The proof is concluded by the application of Theorem 4.2.7. ∎

Notice that the application of the methods of the topological degree theory for multimaps allows one to obtain (under some additional assumptions) the existence of positive equilibrium prices in the given model (see [326]).

Bibliographical comments and additions

Once I got a letter from my reader: "To understand your 'Un-
kept Thoughts' one needs to be very well-read". I telegraphed
him immediately: "You bet!"

—Stanislaw Jerzy Lec

Chapter 1.

The notion of a continuous multivalued map appeared for the first time, as
far to our knowledge, in the paper of a well-known Polish mathematician
K. Kuratowski [269] in 1931 (see also [270]), although in the XIX century
the mathematical analysis operated with multivalued analytical functions.
However, the Kuratowski's definition leaned upon a solid base, the concept
of a topological space elaborated at the beginning of the XX century, and
since that time it began to be studied from various points of view and found
numerous applications.

Among monographs (or large papers of survey type) completely or par-
tially devoted to the theory of multivalued maps, let us distinguish the
works of J. Andres and L. Górniewicz [2], J. Appell, E. De Pascale, H.T.
Nguyen and P.P. Zabreiko [11]; A.V. Arutyunov and V. Obukhovskii [21];
J.-P. Aubin [24], [23]; J.-P. Aubin and A. Cellina [25]; J.-P. Aubin and I.
Ekeland [26]; J.-P. Aubin and H. Frankowska [27]; C. Berge [53], [54]; V.I.
Blagodatskikh [56]; N.A. Bobylev, S.V. Emel'yanov and S.K. Korovin [62];
Yu. G. Borisovich, B.D. Gelman, A.D. Myshkis and V.V. Obukhovskii [70]
– [74]; C. Castaing and M. Valadier [97]; R. Cross [111]; K. Deimling [115],
[116]; Z. Denkowski, S. Migórski and N.S. Papageorgiou [120]; V.V. Fe-
dorchuk and V.V. Filippov [135]; A. Fryszkowski [143]; B.D. Gelman and
V.V. Obukhovskii [175]; L. Górniewicz [190]; S. Hu and N.S. Papageorgiou

[210], [212]; M. Kamenskii, V. Obukhovskii and P. Zecca [235]; E. Klein and A.C. Thompson [246]; W. Kryszewski [264], [265]; J.M. Lasry and R. Robert [279]; B.S. Mordukhovich [310], [311]; V.I. Opoitsev [346]; A. Petruşel and G. Moţ [353]; E.S. Polovinkin [362]; B.N. Pshenichnyi [367]; M. Srebrny [388]; A.A. Tolstonogov [392]; E. Zeidler [414].

Main directions in the development of the theory of multivalued maps and differential inclusions as well as their applications (up to the beginning of the nineties) are reflected in the surveys [70] – [73], [175]. Notice that the total bibliography of these surveys contains more than 2500 items. The detailed exposition of various aspects of the theory and applications with extensive bibliography is contained in the two-volume "Handbook of Multivalued Analysis" [210], [212].

From the very beginning, it should be mentioned that by the present moment the majority of parts of the theory of multivalued maps and differential inclusions are developed so widely that the bibliography presented in this book as well as these comments in no way claim to be complete.

The investigations of the continuity of multimaps and the properties of operations over them arise to the works of K. Kuratowski [269], [270] and C. Berge [53], [54]. (Let us mention, however, the non-correctness of Theorem 13 in §9 [53] and Theorem 3 in §2 Ch. 6 [54] on the intersection of lower semicontinuous multimaps). The study of continuity properties for the intersection of lower semicontinuous multimaps was given in the PhD dissertation of V.V. Obukhovskii (1974), where the notion of the quasi-open multimap was introduced and Theorems 1.3.8 – 1.3.10 were proved. (These results are published in the survey [71]). The existence of a "rectangular" neighborhood of the product of compact sets which can be inscribed into its arbitrary neighborhood which was used in the proof of Theorem 1.3.17 represents the contents of the theorem of Wallace (see [241]). The theorem of maximum (Theorem 1.3.29) was proved by C. Berge [54].

The problem on the existence of a continuous selection for a multimap which goes back to the classical Michael theorem [306] received a wide development and found numerous applications in various branches of mathematics. Mention here the survey [307], the monographs [351], [368], [380] which are specially devoted to this question and also the works [7], [13], [83], [91], [92], [93], [142], [186], [298], [308], [379].

First statements on the existence of continuous single-valued approximations for upper semicontinuous convex-valued multimaps are due to the works of J. von Neumann [318] and S. Kakutani [220]. The approximation result (Theorem 1.4.12) was proved by A. Lasota and Z. Opial [276]. Its particular cases were considered by A. Cellina [98], [99], [100], Yu.G. Borisovich, B.D. Gelman, E. Mukhamadiev and V.V. Obukhovskii [68], [69].

The existence of single-valued approximations for multimaps with contractible values was studied in the series of works of G. Anichini, G. Conti and P. Zecca [5], [6], [7], and for multimaps whose values are absolute retracts this problem was considered by G. Anichini [4].

The results on the approximation of multimaps with aspheric values which are due to the work of A.D. Myshkis [314] found important applications in the topological degree theory and in the theory of differential inclusions. Let us give some explanation. A nonempty compact subset M of a metric space Y is called *aspheric* (or *∞-proximally connected*) if for every $\varepsilon > 0$ there exists $\delta > 0$ such that each continuous map of a unit sphere $\sigma : S^n \to U_\delta(M)$, $n = 0, 1, 2, \ldots$ can be extended to a continuous map of the ball $\tilde{\sigma} : B^{n+1} \to U_\varepsilon(M)$. Further, recall (see, for example, [77]) that a metric space X is *an absolute neighborhood retract* or *an ANR-space*, if for every homeomorphism h that maps it onto a closed subset $h(X)$ of a metric space Y, the set $h(X)$ is a retract of its certain neighborhood. Notice that the class of ANR-spaces is sufficiently large: in particular, a finite-dimensional compact space is an ANR-space if and only if it is locally connected. This means that compact polyhedrons and compact finite-dimensional manifolds are ANR-spaces. The union of a finite number of convex closed subsets of a normed space is also an ANR-space. Now let us mention (see, for example, [190]) that any compact convex or, more generally, contractible subsets of an ANR-space X are aspheric. Moreover, if a subset of such a space X is *an R_δ-set,* i.e., it can be represented as the intersection of a decreasing sequence of compact contractible sets, then it is also aspheric.

A.D. Myshkis' construction of approximation was modified in the paper of Yu.G. Borisovich and Yu.E. Gliklikh [75] as well as in the works of Polish mathematicians. In particular, in the paper of L. Górniewicz, A. Granas and W. Kryszewski [191] (see also [190]) the following assertion was proved: *if X is a compact ANR-space and Y is a metric space then every u.s.c. multimap $F : X \to K(Y)$ with aspheric values admits an ε-approximation for each $\varepsilon > 0$. Moreover, for every $\delta > 0$ there exists such $\varepsilon_0 > 0$ that each two ε-approximations of F $(0 < \varepsilon \leq \varepsilon_0)$ can be joined by a homotopy in the class of δ-approximations.*

Let us mention the work [149] (see also [70]) in which, basing on Myshkis' method, the construction of an obstacle to the existence of a single-valued approximation for a certain class of multimaps was suggested. The questions of characterization of multimaps by single-valued approximations are considered in the works [58], [61].

The connection between continuous selection and approximation results for multimaps is studied in the works [165], [176] and [190].

The systematic study of measurable multifunctions started in the sixties when their interesting applications to theory of differential inclusions, problems of optimal control, convex analysis and other branches of mathematics were revealed. Among the first works of this direction, let us mention the investigations of C. Castaing [95], G. Debreu [113], M.Q. Jacobs [219], C. Olech [344], A. Pliś [357], R.T. Rockafellar [372]. A detailed exposition of the theory of measurable multifunctions can be found in the monograph of C. Castaing and M. Valadier [97]. A large number of questions concerning theory and applications of measurable multifunctions are described also in monographs and works [11], [12], [25], [27], [56], [71], [72], [74], [115], [116], [175], [184], [195], [205], [206], [207], [208], [210], [215], [216], [235], [245], [246], [310], [373], [388], [392], [402].

Various versions of Theorem 1.5.6 can be found either in book [97] or in many of the above mentioned works. It is worth noting that the existence of a measurable selection was proved by K. Kuratowski and C. Ryll-Nardzewski in [271]. An earlier version of this theorem is contained in the work of V.A. Rokhlin [374]. The existence of the Castaing representation is proved in [95]. Notice however that the construction of a dense "tube" of continuous selections for a l.s.c. multimap is due to E. Michael [306]. The Lusin property for multimaps was studied by A. Pliś [357] and M.Q. Jacobs [219].

The definition of the multivalued integral which we present is due to R. Aumann [29]. Another approaches were suggested, for example, by G. Debreu [113] and M. Hukuhara [214]. The proof of Theorem 1.5.12 as well as the description of other properties of the multivalued integral can be found in the books of J.-P. Aubin and H. Frankowska [27], B. Glodde and H.-D. Niepage [184], W. Hildenbrand [206], S. Hu and N.S. Papageorgiou [210], A.D. Ioffe and V.M. Tikhomirov [216], E. Klein and A.C. Thompson [246], B.S. Mordukhovich [310], E.S. Polovinkin [362], A.A. Tolstonogov [392] and others. A detailed bibliography on the multivalued integral is presented in surveys [71], [175].

Properties of the Carathéodory type multifunctions and the superposition multioperator are described in monographs [11], [74], [116], [210],

[212], [235], [392]. Theorem 1.5.14 for the case of finite-dimensional spaces was proved by N.Kikuchi [244]. Counterexamples 1.5.20 (a, b) are constructed by V.V. Obukhovskii (see [324] and [10]). Theorem 1.5.22 on a superpositionally selectable multimap was proved by C. Castaing [96]. The property of the closedness of the superposition multioperator was studied by A. Lasota and Z. Opial [274].

Starting with the paper [137], various versions of the Filippov lemma were considered in many papers. Let us mention some of them: [95], [130], [195], [208], [218], [310], [402]. (A more detailed bibliography can be found in surveys [71] and [175]). A complete proof of Theorem 1.5.36 on the lower semicontinuity of the superposition multioperator can be found in [116], Theorem 9.3.

Let us mention that such an important part of the contemporary multivalued analysis as the differential calculus of multimaps remained outside the scope of our exposition. The description of various approaches to the definition of the concept of a derivative for a multifunction, some applications and bibliography can be found in the monographs [27], [210], [386] and the surveys [71], [175].

Chapter 2.

Beginning with the papers of S.B. Nadler, Jr [315] and J.T. Markin [305], there is a very large number of works devoted to fixed point theorems for contractive and nonexpansive multimaps (the bibliography up to 1987 is given in the surveys [71] and [73]). Mention the exposition of these questions in the monographs [185], [190], [198], [210], [245], [353], [378], [406] and others. Theorem 2.1.2 was proved by Nadler [315] for the case of a multimap with closed bounded values and was extended in the work [110] to the case of closed-valued multimaps. Theorem 2.1.3 is given in the paper [383]. In the same work some other results on the structure of the fixed point set of multivalued contractions are obtained. The topological dimension of the fixed point set for multivalued contractive maps was studied in [126]. Theorem 2.1.4 was proved by B.D. Gelman. A local version of this result is published in the paper [160]. Theorem 2.1.6 is a version of B. Ricceri's theorem [369]. The technique of the theory of multivalued maps was applied for the first time to the study of equations with closed surjective linear operators in the work [164]. On some other results of this direction see [152], [159], [161], [163], [169], [171], [178]. Theorem 2.1.11 for the case of a continuous surjective operator was proved in [369]. Corollary 2.1.13 is due to the works of I.C. Gohberg and M.G. Krein in which it was proved by other methods. Let us mention the works [166], [167], where the applications to degenerate differential equations in a Banach spaces are considered. The exposition of inequalities of Caristi type and α-contractive multimaps is based on the paper of B.D. Gelman [173].

In the paper of A.V. Arutyunov [14] (see also [21]) for a pair of multimaps, one of which is α-covering whereas the other one is Lipschitz some theorems on a coincidence point are obtained. The stability of such coincidence points is studied in the work [15] (see also [21]). Local versions of these results are contained in the paper [17], where applications to the existence of positive trajectories for a semilinear control system are considered. Coincidence points of Lipschitz and covering tuples of multimaps are studied in [172]. The paper of A.V. Arutyunov and B.D. Gelman [18] in which a theorem on a minimum of a functional on a metric space is used for the proof of fixed point theorems for multimaps is close to these works. In the paper [20] a theorem on a coincidence point for a pair of h-upper semicontinuous multimaps with convex closed values in a finite-dimensional space is presented.

First constructions of the topological degree for convex-valued multifields in Banach spaces were suggested by A. Granas in 1959 in the works [196], [197]. However the technique applied by Granas used "too strong weapon": homological methods going back to the known work of S. Eilenberg and D. Montgomery [127] and based on the Vietoris–Begle theorem on isomorphism (see, for example, [387]). The application of the method of single-valued approximations for the construction of the degree theory is much more simple, geometrically clear and often allows to reduce the evaluation of the degree to the corresponding "single-valued" situations. Such an approach was suggested simultaneously and independently by Yu.G. Borisovich, B.D. Gelman, E. Mukhamadiev, V.V. Obukhovskii [68], [69] and A. Cellina and A. Lasota [101]. Let us mention also the work of M. Hukuhara [213] which defined the topological degree of multifields with the help of multivalued continuous approximations. Based on approximation methods, the systematic construction of the topological degree theory in locally convex topological vector spaces for compact multifields and its applications to fixed point theorems is contained in the works of Yu.G. Borisovich, B.D. Gelman, A.D. Myshkis, V.V. Obukhovskii [70] (the relative case) and T.-W. Ma [297] (the "absolute" case).

The extension of the method of single-valued approximations to the case of nonconvex-valued multimaps made it possible to give it the character of a fairly universal tool for the construction of the topological degree for a great variety of classes of multimaps (see monographs [2], [190], [210], [264] and the works [37], [59], [191], [253], [340]).

At the same time the homological methods proved their high efficiency in the problem of construction of the topological degree for multimaps with acyclic values, their compositions and multimaps of a more complicated structure. (Recall that a set is called *acyclic* if it has the same homologies as a singleton. Convex, contractible, R_δ-sets are acyclic).

Homological methods in the topological degree and fixed point theory of multimaps are studied and described in the surveys [70] – [73], monographs [2], [125], [177], [190], [235], [264], [265], [279], works [88], [106], [144], [153], [192], [199], [267], [268], [277], [327], [407].

Let us mention the papers of B.D. Gelman [150], [151], [153] and the works [51], [52] in which topological characteristics of multimaps being generalizations of the topological degree were studied. These constructions were based on the study of a homological structure of graphs of multimaps.

One more interesting and intensively developing direction in the topological degree theory for multimaps is connected with its extension to the case of noncompact (condensing, ultimately compact, fundamentally restrictible) multifields in infinite-dimensional spaces. In this connection, the class of condensing multimaps attracted the greatest attention that was caused on the one hand by the fact that this class is a very natural enlargement of the collection of compact multimaps and on the other hand the multimaps of this type can be effectively used in the theory of differential equations and inclusions in Banach spaces (see below).

The most important property of a condensing multimap is the existence of an invariant convex closed subset (*a fundamental set*) a priori containing all fixed points and possessing the compact restriction property. This makes it possible to apply the theory of relative topological degree for compact multifields for the construction of the topological degree of condensing multifields. Such approach was applied in the paper of V.V. Obukhovskii [322], where the construction of a topological degree for a condensing multifield in a Frechét space was suggested. In the sequel, various methods of the construction of the topological degree of condensing and other noncompact multifields were described in the works of W.V. Petryshyn and P.M. Fitzpatrick [354], J.R.L. Webb [405], A. Vanderbauwhede [400]. The topological degree theory was extended to the case of condensing acyclic multimaps in the work of V.V. Obukhovskii [327] (see also the survey [72] and the works of L. Górniewicz and Z. Kucharski [192]; G. Conti, V. Obukhovskii and P. Zecca [106]). The degree for condensing multimaps which can be represented as a composition of a superposition multioperator with an abstract solving operator is defined in the paper of R. Bader, M.I. Kamenskii and V.V. Obukhovskii [36]. A systematic exposition of the topological degree theory for various classes of noncompact multimaps (with additional references therein) can be found in the surveys [70], [71], [72] and in the monograph of M. Kamenskii, V. Obukhovskii and P. Zecca [235].

An important and finding interesting applications direction in the topological degree theory is the development of *the coincidence degree* for various types of multimaps with linear and nonlinear operators. Let us mention here the works of Yu.G. Borisovich, [66], [67], E. Tarafdar and S.K. Teo [389], T. Pruszko [366], D. Gabor and W. Kryszewski [145], V. Obukhovskii, P. Zecca and V. Zvyagin [340], [342], S.V. Kornev and V.V. Obukhovskii [253] (see also monographs [2], [116], [210], [265], [339], [401]). In the work of V. Obukhovskii, P. Zecca and V. Zvyagin [343] the coincidence degree of a linear Fredholm operator with a compact multimap is used in Landesman–Laser type theorems on the solvability of equations containing a Fredholm operator with a nontrivial kernel and a discontinuous nonlinearity. An application of the coincidence degree to the optimization of an impulsive control system is studied in the paper of Y.-C. Liou, V. Obukhovskii and J.-C. Yao [280].

The index of solvability for variational inequalities with a multioperator in a Banach space was constructed in the paper of I. Benedetti and V. Obukhovskii [45]. This degree type topological characteristic is used to obtain some results on the existence of solutions to variational inequalities. As applications, the minimization of a nonsmooth functional and a problem of the mathematical economics are considered. In the paper of Y.-C. Liou, V. Obukhovskii and J.-C. Yao [290] the notion of an index of solvability for a variational inequality on a Riemannian manifold involving a multivalued vector field is introduced and investigated.

Applications of the topological degree methods to the study of *bifurcations of solutions* to operator inclusions are given in the works [286], [290], [291] (see also the monographs [190], [265], [339]).

The problem on topological properties of a fixed point set arises very naturally in various aspects of the theory of multimaps. The connectedness and acyclicity of a fixed point set were studied in the works [2], [146], [152], [155], [162], [190], [278], [279] and others. The questions of the topological dimension of the fixed point and solution sets are described in [35], [126], [152], [155], [382], [383] and other works. In the paper [19] the structure of the set of coincidence points for two multimaps of metric spaces is studied.

It is worth noting that fixed point results for multimaps of various types, constructions of topological invariants different from the degree (the Lefschetz and Nielsen numbers), problems on the solvability of operator inclusions, their applications and other questions are contained in the works [2], [24], [26], [25], [27], [33], [50], [66], [67], [70], [75], [86], [87], [107], [115], [116], [117], [127], [128], [133], [147], [154], [185], [190], [198], [203], [210], [325], [353], [378], [409], [414].

In the papers [3], [390] (see also [190]) the topological degree and the coincidence degree were defined for multimaps depending on a random parameter. The construction of the topological fixed point index for a class of random condensing compositions of multimaps with aspheric values is presented in the paper [180]. A random version of a theorem on equilibrium points for two parametrized multivalued maps satisfying a joint Caristi type condition is given in [179].

Notice that in the paper [201] the connection between fixed point theorems for multimaps of various types, the Knaster–Kuratowski–Mazurkiewicz property, variational inequalities, minimax relations and the Gale–Nikaido–Debreu theorem on excess demand is studied. The stability of a fixed point in terms of its topological index was considered in the paper [332]. Let us mention the direction connected with the computation of fixed points: [391]. Applications of fixed points and the technique of the theory of multivalued maps to the obstacle problems, free boundary problems and problems arising in the theory of thermostats are described in the works [102], [103], [181]. Let us point again to surveys [71], [72], [73], [175], where an additional information and references should be found.

Chapter 3.

The rapid development of the theory of differential inclusions, which started in the sixties of the last century continues at the present time as well. A large number of its parts which are traditional in the general theory of differential equations (existence of solutions, their qualitative properties, including the topological structure of the solution sets, continuous dependence on initial data and parameters, stability, existence of periodic solutions etc.) are elaborated. At the same time some interesting specific problems were detected. They include, for example, the question on relations between solution sets of an initial and "convexified" problems, the description of control systems by differential inclusions and other ones.

At the present time there are several sufficiently informative and detailed monographs completely or partially presenting various aspects of the theory of differential inclusions and its applications. We can refer to the following sources: J. Andres and L. Górniewicz [2]; A.V. Arutyunov and V. Obukhovskii [21]; J.-P. Aubin [23]; J.-P. Aubin and A. Cellina [25]; J.-P. Aubin and H. Frankowska [27]; M. Benchohra and S. Abbas [44], C. Castaing and M. Valadier [97]; F. Clarke [104]; K. Deimling [116]; O.P. Filatov and M.M. Khapaev [136]; A.F. Filippov [141]; L. Górniewicz [190]; S. Hu and N.S. Papageorgiou [211], [212]; M. Kamenskii, V. Obukhovskii and P. Zecca [235]; M. Kisielewicz [245]; B.S. Mordukhovich [310], [312]; N.A. Perestyuk, V.A. Plotnikov, A.M. Samoilenko, N.V. Skripnik [352], V.A. Plotnikov, A.V. Plotnikov and A.N. Vityuk [361]; G.V. Smirnov [386]; A.A. Tolstonogov [392], Y. Zhou [418]. The additional bibliography and the survey of various research directions can be found also in surveys [71] and [175].

In the sixties various existence theorems for differential inclusions with a convex-valued right-hand side in a finite-dimensional space were obtained in the works of C. Castaing [96], J.L. Davy [112], N. Kikuchi [242], [243], [244], A. Lasota and Z. Opial [274], A. Pliś [358] – [360] and others. It should be mentioned that apparently for the first time in the work [274] the topological methods (the Bohnenblust–Karlin fixed point theorem) were applied.

For nonconvex-valued differential inclusions the first existence theorem was proved by A.F. Filippov [139] under assumption that the right-hand side satisfies the Lipschitz condition. Then he obtained also the existence result for differential inclusions with continuous right-hand side [140]. This result was generalized in the work of C. Olech [345]. For differential inclusions with a l.s.c. nonconvex-valued right-hand side the first results were obtained by A. Bressan [78] and S. Lojasiewicz [295]. In the sequel, *the method of directionally continuous selections* elaborated by Bressan provided an opportunity to develop a unified approach to the study of differential inclusions with a u.s.c. and l.s.c. right-hand part (see [79] – [82]).

The connectedness of the solution set in the case when the right-hand part of an inclusion satisfies the upper Carathéodory conditions was established by J.L. Davy [112]. (Notice that the connectedness of the attainability set was proved earlier by N. Kikuchi [243]). In the sequel it was discovered that the solution set in this case has a more refined topological structure: J.-M. Lasry and R. Robert [278] proved its acyclicity and later it was demonstrated that it is an R_δ-set (C.J. Himmelberg and F.S. Van Vleck [209] in the case of a u.s.c. right-hand side and F.S. De Blasi and J. Myjak [60] in a general case). In the sequel this result was developed in various directions: to the case of inclusions with constraints, inclusions in a Banach space, fractional-order inclusions etc. We refer to monographs [2], [116], [123], [190], [212], [235], and papers [32], [38], [105], [229], [266], [356], [394], [396]. It is worth noting that the R_δ-property of the solution set plays a very important role in the investigation of a periodic and more general boundary value problems for differential inclusions: by virtue of the above mentioned approximability of multimaps with R_δ-values, it allows to apply the topological degree and fixed point methods to solving of these problems (let us add to the above references the following items: [31], [59], [194], [226], [233], [333], [355], [230]).

The property of connectedness of the solution set for a differential inclusion with a l.s.c. right-hand side was established by A. Bressan [79] (see also [116]). For the case of inclusions in a Banach space this property was investigated by A. Bressan and V. Staicu [84] and V. Obukhovskii and P. Zecca [334] (see also [235]).

The topological dimension of the solution set for the Cauchy problem was studied in the works [126], [155], [156]. [158]. Generalizations of these results to the cases of inclusions of a more general form and more general problems were obtained in the papers [28], [34] and others.

Some properties of solution sets of differential equations and inclusions are described in the monograph of R. Dragoni, J.W. Macki, P. Nistri and P. Zecca [123].

In the recent decades the differential inclusions in Banach spaces are studied very intensively. This is largely due to the fact that inclusions of that type find interesting and important applications in the investigations of control processes governed by partial differential equations. Differential inclusions in Banach spaces are described in the monographs of K. Deimling [116], S. Hu and N.S. Papageorgiou [211], [212], M. Kamenskii, V. Obukhovskii and P. Zecca [235], B.S. Mordukhovich [312], A.A. Tolstonogov [392] and in a large number of papers (see, for example, [8], [31], [32], [34], [36], [38], [40], [105], [122], [187], [188], [193], [221] – [234], [266], [300], [328], [331], [333] – [335], [347] – [350], [393] – [398], [413]). Notice that in the monograph [235] an approach to the investigation of semilinear differential inclusions with a unbounded linear operator in Banach spaces based on the topological degree theory for condensing multimaps is systematically developed. Semilinear differential inclusions in Banach spaces with a nondensely defined linear part were studied in the paper [337]. The investigation of functional inclusions with causal operators in Banach spaces and some of their applications was realized in [338]. The same topological methods are applied in the paper [239] to the study of an abstract Cauchy problem for higher order functional differential inclusions with infinite delay in a Banach space.

Second-order differential inclusions in finite-dimensional and Hilbert spaces with some applications to control systems and mathematical physics were considered in the paper of N.V. Loi and V. Obukhovskii [288] and monograph [339].

In the theory of *degenerate differential equations and inclusions* in Banach spaces the methods connected with the employment of *multivalued linear operators* found very effective applications (see, for example, the monographs of R. Cross [111], A. Favini and A. Yagi [134] and the papers [335], [40]), [9], [281]). In the paper of A. Baskakov, V. Obukhovskii and P. Zecca [41] the solvability of differential inclusions with multivalued linear operators in homogeneous spaces of functions (including the spaces of periodic and almost periodic functions) is investigated, applications to control systems governed by degenerate differential equations are given. In the paper [42] almost periodic solutions at infinity for differential equations and inclusions are studied. Existence of local solutions and properties of the solution set for a degenerate differential inclusion are studied also in the paper of B.D. Gel'man [168].

In the paper [294] a nonlocal problem for projected differential equations and inclusions in finite dimensional spaces is considered. It is shown how the abstract results can be applied to the study of a market model with the price intervention in the form of price floors and ceilings.

The investigation of periodic solutions to differential equations goes back apparently to the work of A. Lasota and Z. Opial [274], where the solvability of a general boundary value problem, including a periodic one, was studied. A method of the investigation of a periodic problem based on the approximation of a differential inclusion by differential equations was suggested by A.I. Povolotskii and E.A. Gango [364], [365]. The method of integral multivalued operators in the periodic problem was developed in the works of B.D. Gelman [148] and V.V. Obukhovskii [323], [324] (see also surveys [70], [72]). Theorem 3.3.10 was proved by B.D. Gelman (see [70], [74]). The method of the multivalued translation operator starts with the works of B.D. Gelman (see the survey [70]), J.-M. Lasry and R. Robert [277], K. Deimling [114]. Its exposition and connection with the method of guiding functions can be found in the monographs of J. Andres and L. Górniewicz [2], L. Górniewicz [190], V. Obukhovskii, P. Zecca, N.V. Loi and S. Kornev [339] and in the works [31], [32], [38], [59], [188], [194], [202], [222], [226], [233], [266], [320], [331], [333], [355] and others. Based on the theory of condensing multimaps, the application of the method of integral multioperators and the method of the translation multioperator to the periodic problem is described in the monograph of M. Kamenskii, V. Obukhovskii and P. Zecca [235]. The topological degree theory for condensing multimaps is applied in the paper of M. Kamenskii, V. Obukhovskii and J.-C. Yao [232] to the study of periodic solutions to a neutral type operator differential inclusion. The study of boundary value problems more general than the periodic one is presented in the works [9], [40], [43], [47], [48], [94], [122], [230], [335], [348], [350], [413] (see also [212], [265]). Bifurcations of families of periodic solutions are described in the monographs [265], [339] and papers [287], [293]. Modern developments of the method of guiding functions and its applications to the periodic and bifurcation problems as well as to asymptotic behavior of solutions can be found in the monograph [339] and the papers [251], [252], [254], [255], [256], [257], [258], [259], [285], [289], [292], [330].

In the recent years the theory of differential equations and inclusions of a fractional order is the subject of an active study for a large number of researchers. One of the main reasons of this interest is caused by important applications of this theory in physics, engineering, biology, economics, and other sciences. It should be mentioned, in this connection, that fractional order models provide an effective and convenient machinery for the description of systems with memory and hereditary properties. The existence of local and global solutions to semilinear fractional-order differential inclusions in a Banach space, their approximation, qualitative properties of solutions, including the topological structure of the solution set and its continuous dependence on initial data and parameters, the solvability of periodic and more general boundary value problems and other questions are studied in the cycle of works of M. Kamenskii, V. Obukhovskii, G. Petrosyan and J.-C. Yao [227] – [231] by the methods of the theory of condensing multimaps. In the papers of I. Benedetti, V. Obukhovskii and V. Taddei [46]– [48] the application of the weak topology methods made it possible to obtain some results concerning inclusions of fractional order without compactness assumptions. For results of this direction see also the papers [200], [237], [240] and the monograph of Y. Zhou [418].

The methods of the theory of differential inclusions are effectively applied to solving various problems of differential variational inequalities (see [237], [283], [284], [296]).

The averaging principle for differential inclusions is exposed in the monograph of O.P. Filatov and M.M. Khapaev [136]. This question is studied also in the book [235] and the works [108], [109], [221].

The viability theory occupies a very important place in the contemporary theory of differential inclusions. The principal problem of this theory is the study of trajectories belonging to a prescribed closed subset of the phase space. Concerning these questions we refer to the monographs [23] – [27], [116], [190], [386]. Among the papers of this direction, let us mention [32], [38], [193], [266], [321], [355], [397], [398].

Differential inclusions on Riemannian manifolds were studied in the works [30], [183].

By virtue of the close connection of the theory of differential inclusions with the theory of control systems and optimization problems, these questions are considered in practically all above mentioned books. Let us indicate the books [245] and [211] specially devoted to this problem. A survey of certain results in this direction can be found in [175]. The application of the topological degree theory for condensing multimaps allows to solve *the controllability problem* for systems governed by semilinear differential inclusions in Banach spaces (see [336], [49], [281]). The controllability result obtained by the weak topology method is described in the paper [46]. A controllability for a system governed by a second-order differential inclusion with a nonlocal condition is studied in the paper of T.D. Ke and V. Obukhovskii [238]. Extremal problems for differential inclusions and *the Pontryagin maximum principle*, are studied in the works of A.V. Arutyunov, S.M. Aseev, V.I. Blagodatskikh [16], S.M. Aseev [22], V.I. Blagodatskikh [55], V.I. Blagodatskikh and A.F. Filippov [57], B.S. Mordukhovich [310], [312], B.N. Pshenichnyi [367]. The survey of topological methods applicable to some optimization problems for systems governed by differential inclusions is given in the paper of V. Obukhovskii [329]. Applications of inclusions with multivalued operators to some optimization problems are considered in the work of V. Zvyagin, V. Obukhovskii and A. Zvyagin [419]. Among other works, let us mention [8], [41], [76], [200], [222], [226], [242], [243], [275], [320], [321], [323], [328], [331], [340], [347] – [350].

The topological degree theory for multivalued maps was applied in the works of V. Obukhovskii, P. Zecca, V. Zvyagin [341] and C. Gori, V. Obukhovskii, P. Rubbioni and V. Zvyagin [189] to the solving of *optimization problems of hydrodynamics*.

A survey of works on *integral inclusions* can be found in [175] (see also the works of A.I. Bulgakov [91]– [93]).

Applications of differential inclusions in the theory of *differential games* are considered in the monographs of N.N. Krasovskii and A.I. Subbotin [263] and J.-P. Aubin [23]. Methods of the theory of differential inclusions are used to the study of systems controlled in conditions of uncertainty (see the book of A.B. Kurzhanskii [272]). M.A. Krasnoselskii and A.V. Pokrovskii [261] applied differential inclusions to the investigation of *systems with hysteresis*, in the book of M.D.P. Monteiro Marques [309] applications to problems of mechanics are described. Some examples of applications to a model of a transmission line with nonlinear units and to a model of a hybrid system with a dry friction are described in the book [235]. The detailed exposition of some problems in the theory of evolution inclusions and variational inequalities and their various applications is presented by M.Z. Zgurovsky, V.S. Mel'nik, P.O. Kasyanov [415], [416] and M.Z. Zgurovsky, P.O. Kasyanov, O.V. Kapustyan, J. Valero, and N.V. Zadoianchuk [417].

Chapter 4.

The notion of a generalized (or *dispersive*) dynamical system arised in connection with the study of ordinary differential equations with a non-unique solution and differential inclusions and it was described in the works of E.A. Barbashin [39], B.M. Budak [89], [90], A.D. Myshkis [314] and others. Let us mention the monograph of K.S. Sibirskii and A.S. Shube [385] in which topological properties of generalized dynamical systems are studied and the cycle of works of E. Roxin [375] – [377] on the stability of generalized dynamical systems. Attractors of generalized dynamical systems are described in the monograph [417]. We give the axiomatics of generalized dynamical systems suggested by Barbashin.

The problem of the study of rest points of generalized dynamical systems was set up by Myshkis in [314]. In the same work the idea to consider one-sided systems satisfying axioms $(G1)$–$(G3)$ was suggested. Presented theorems on the existence of rest points are particular cases of assertions from the paper of V. Obukhovskii [325]. Various results on the existence of rest points were proved also in works [314], [70], [157] and others. Let us mention the work [157] in which a class of *approximable* generalized dynamical systems was indicated.

As it was mentioned already, problems of the game theory and mathematical economics were among first objects of applications for ideas and methods of the theory of multivalued maps. And in fact, already pioneering works on fixed point principles for multimaps contained, as direct consequences, one or other theorems on the existence of optimal strategies (see [63], [132]), [182], [220], [318]). Applications of the technique of multivalued maps and differential inclusions to the theory of games and mathematical economics are described in the monographs of J.-P. Aubin [24]; J.-P. Aubin and A. Cellina [25]; J.-P. Aubin and I. Ekeland [26]; C. Berge [53], [54]; Z. Denkowski, S. Migórski and N.S. Papageorgiou [121]; W. Hildenbrand [206]; S. Hu and N.S. Papageorgiou [212]; S. Karlin [236]; E. Klein and A.C. Thompson [246]; C. Lancaster [273]; V.L. Makarov and A.M. Rubinov [299]; B.S. Mordukhovich [312]; H. Nikaido [319]; M.J. Todd [391]; K. Urai [399], G.X.-Z. Yuan [409]. The surveys of the literature can be found in [70], [71], [175].

Bibliography

Then, catching sight of the titles here, he began to breathe easily and with an audible sighing of relief. He was among friends.

—*Raymond F. Jones* "Noise level"

[1] P.S. Alexandroff. *Einführung in die Mengenlehre und in die allgemeine Topologie.* Hochschulbucher für Mathematik, 85. VEB Deutscher Verlag der Wissenschaften, Berlin, 1984.

[2] J. Andres, L. Górniewicz. *Topological Fixed Point Principles for Boundary Value Problems.* Kluwer, Dordrecht, 2003.

[3] J. Andres, L. Górniewicz. Random topological degree and random differential inclusions. *Topol. Methods Nonlinear Anal.* 40 (2012), no. 2, 337–358.

[4] G. Anichini. Approximate selections for nonconvex set valued mappings. *Boll. Un. Mat. Ital.* B (7) 4 (1990), no. 2, 313–326.

[5] G. Anichini, G. Conti, P. Zecca. Approximation of nonconvex set valued mappings. *Boll. Un. Mat. Ital.* C (6) 4 (1985), no. 1, 145–154.

[6] G. Anichini, G. Conti, P. Zecca. A further result on the approximation of nonconvex set valued mappings. *Boll. Un. Mat. Ital.* C (6) 4 (1985), no. 1, 155–171.

[7] G. Anichini, G. Conti, P. Zecca. Approximation and selection for nonconvex multifunctions in infinite-dimensional spaces. *Boll. Un. Mat. Ital.* B (7) 4 (1990), no. 2, 410–422.

[8] G. Anichini, P. Zecca. Multivalued differential equations in Banach space. An application to control theory. *J. Optimization Theory Appl.* 21 (1977), no. 4, 477–486.

[9] Q.H. Ansari, Y.C. Liou, V. Obukhovskii, N.C. Wong. Topological degree methods in boundary value problems for degenerate functional differential inclusions with infinite delay. *Taiwanese J. Math.* 12 (2008), no. 7, 1827–1847.

[10] J. Appell. Multifunctions of two variables: examples and counterexamples. *Topology in nonlinear analysis (Warsaw, 1994)*, Banach Center Publ., 35, Polish Acad. Sci., Warsaw, 1996, 119–128.

[11] J. Appell, E. De Pascale, H.T. Nguyen, P.P. Zabreiko. Multi-valued superpositions. *Dissertationes Math. (Rozprawy Mat.)* 345 (1995), 1–97.

[12] V.I. Arkin, V.L. Levin. Convexity of values of vector integrals, measurable choice theorems and variational problems *Russian Math. Surveys.* 27 (1972), no. 3, 21–85.

[13] A.V. Arutyunov. Specific selections for set-valued mappings. *Doklady Mathematics,* 63 (2001), no. 2, 182–184.

[14] A.V. Arutyunov. Covering mappings in metric spaces and fixed points. *Doklady Mathematics,* 76 (2007), no. 2, 665–668.

[15] A.V. Arutyunov. Stability of coincidence points and properties of covering mappings. *Mathematical Notes* 86 (2009), no. 1–2, 153–158.

[16] A.V. Arutyunov, S.M. Aseev, V.I. Blagodatskikh. First-order necessary conditions in the problem of optimal-control of a differential inclusion with phase constraints. *Sbornik Mathematics* 79 (1994), no. 1, 117–139.

[17] A. Arutyunov, E. Avakov, B. Gel'man, A. Dmitruk, V. Obukhovskii. Locally covering maps in metric spaces and coincidence points. *J. Fixed Point Theory Appl.* 5 (2009), no. 1, 105–127.

[18] A.V. Arutyunov, B.D. Gel'man. Minimum of a functional in a metric space and fixed points. *Comput. Math. Math. Phys.* 49 (2009), no. 7, 1111–1118.

[19] A.V. Arutyunov, B.D. Gel'man. On the structure of the set of coincidence points. *Sb. Math.* 206 (2015), no. 3, 370–388.

[20] A. Arutyunov, B. Gelman, V. Obukhovskii. A coincidence theorem for multivalued maps and its applications. *J. Fixed Point Theory Appl.* 17 (2015), no. 2, 331–340.

[21] A.V. Arutyunov, V. Obukhovskii. *Convex and Set-Valued Analysis. Selected Topics.* De Gruyter Graduate. Walter de Gruyter, Berlin-Boston, 2017.

[22] S.M. Aseev. Extremal problems for differential inclusions with state constraints. *Proc. Steklov Inst. Math.* 233 (2001), 1–63.

[23] J.-P. Aubin. *Viability Theory.* Birkhäuser, Boston-Basel-Berlin, 1991.

[24] J.-P. Aubin. *Optima and Equilibria. An Introduction to Nonlinear Analysis.* Springer-Verlag, Berlin-Heidelberg-New York, 1993.

[25] J.-P. Aubin, A. Cellina. *Differential Inclusions. Set-Valued Maps and Viability Theory.* Springer-Verlag, Berlin-Heidelberg-New York-Tokyo, 1984.

[26] J.-P. Aubin and I. Ekeland. *Applied Nonlinear Analysis.* Wiley-Interscience, New York, 1984.

[27] J.-P. Aubin, H. Frankowska. *Set-Valued Analysis.* Birkhäuser, Boston-Basel-Berlin, 1990.

[28] A. Augustynowicz, Z. Dzedzej, B.D. Gelman. The solution set to BVP for some functional-differential inclusions. *Set-Valued Anal.* 6 (1998), no. 3, 257–263.

[29] R.J. Aumann. Integrals of set-valued functions. *J. Math. Anal. Appl.* 12 (1965), 1–12.

[30] S.V. Azarina, Yu.E. Gliklikh, A.V. Obukhovskii. Solvability of Langevin differential inclusions with set-valued diffusion terms on Riemannian manifolds. *Appl. Anal.* 86 (2007), no. 9, 1105–1116.

[31] R. Bader. The periodic problem for semilinear differential inclusions in Banach spaces. *Comment. Math. Univ. Carolin.* 39 (1998), no. 4, 671–684.

[32] R. Bader. On the semilinear multi-valued flow under constraints and the periodic problem. *Comment. Math. Univ. Carolin.* 41 (2000), no. 4, 719–734.

[33] R. Bader, G. Gabor, W. Kryszewski. On the extension of approximations for set-valued maps and the repulsive fixed points. *Boll. Un. Mat. Ital.* B (7), 10 (1996), no. 2, 399–416.

[34] R. Bader, B.D. Gel'man, M. Kamenskii, V. Obukhovskii. On the topological dimension of the solutions sets for some classes of operator and differential inclusions. *Discuss. Math. Differ. Incl. Control Optim.* 22 (2002), no. 1, 17–32.

[35] R. Bader, B.D. Gel'man, V.V. Obukhovskii. On a class of multivalued maps. *Vestnik Voronezh. Gos. Univ. Ser. Phys.–Mat.* 2 (2003), 35–38 (in Russian).

[36] R. Bader, M. Kamenskii, V. Obukhovskii. On some classes of operator inclusions with lower semicontinuous nonlinearities. *Topol. Methods Nonlinear Anal.* 17 (2001), no. 1, 143–156.

[37] R. Bader, W. Kryszewski. Fixed-point index for compositions of set-valued maps with proximally ∞-connected values on arbitrary ANR's. *Set-Valued Anal.* 2 (1994), no. 3, 459–480.

[38] R. Bader, W. Kryszewski. On the solution sets of differential inclusions and the periodic problem in Banach spaces. *Nonlinear Anal.* 54 (2003), no. 4, 707–754.

[39] E.A. Barbashin. On the theory of generalized dynamical systems. *Uchen. Zap. Moscow Univ., Matematika.* 135 (1948), no. 2, 110–113 (in Russian).

[40] A. Baskakov, V. Obukhovskii, P. Zecca. Multivalued linear operators and differential inclusions in Banach spaces. *Discuss. Math. Differ. Incl. Control Optim.* 23 (2003), 53–74.

[41] A. Baskakov, V. Obukhovskii, P. Zecca. On solutions of differential inclusions in homogeneous spaces of functions. *J. Math. Anal. Appl.* 324 (2006), no. 2, 1310–1323.

[42] A. Baskakov, V. Obukhovskii, P. Zecca. Almost periodic solutions at infinity of differential equations and inclusions. *J. Math. Anal. Appl.* 462 (2018), no. 1, 747–763.

[43] M.M. Basova, V.V. Obukhovskii. On some boundary value problems for functional-differential inclusions in Banach spaces. *J. Math. Sci. (N.Y.)* 149 (2008), no. 4, 1376–1384.

[44] M. Benchohra, S. Abbas. *Advanced Functional Evolution Equations and Inclusions.* Developments in Mathematics, 39. Springer, Cham, 2015.

[45] I. Benedetti, V. Obukhovskii. On the index of solvability for variational inequalities in Banach spaces. *Set-Valued Anal.* 16 (2008), no. 1, 67–92.

[46] I. Benedetti, V. Obukhovskii, V. Taddei. Controllability for systems governed by semilinear evolution inclusions without compactness. *NoDEA Nonlinear Differential Equations Appl.* 21 (2014), no. 6, 795–812.

[47] I. Benedetti, V. Obukhovskii, V. Taddei. On noncompact fractional order differential inclusions with generalized boundary condition and impulses in a Banach space. *J. Funct. Spaces.* 2015, Art. ID 651359, 10 pp.

[48] I. Benedetti, V. Obukhovskii, V. Taddei. On generalized boundary value problems for a class of fractional differential inclusions. *Fract. Calc. Appl. Anal.* 20 (2017), no. 6, 1424–1446.

[49] I. Benedetti, V. Obukhovskii, P. Zecca. Controllability for impulsive semilinear functional differential inclusions with a non-compact evolution operator. *Discuss. Math. Differ. Incl. Control Optim.* 31 (2011), no. 1, 39–69.

[50] H. Ben-El-Mechaiekh, W. Kryszewski. Equilibria of set-valued maps on nonconvex domains. *Trans. Amer. Math. Soc.* 349 (1997), no. 10, 4159–4179.

[51] N.M. Benkafadar, B.D. Gel'man. Generalized local degree for multi-valued mappings. *Internat. J. of Math., Game Theory and Algebra.* 10(2000), no. 5, 413–434.

[52] N.M. Benkafadar, B.D. Gel'man. On some topological invariants for morphisms defined in homological spheres. *Applied General Topology.* 16 (2015), no. 1, 19–30.

[53] C. Berge. *Théorie Générale des Jeux à n Personnes.* Mémor. Sci. Math., 138. Gauthier-Villars, Paris, 1957.

[54] C. Berge. *Espaces Topologiques. Fonctions Multivoques.* 2nd ed., Dunod, Paris, 1966.

[55] V.I. Blagodatskikh. The maximum principle for differential inclusions *Proc. Steklov Inst. Math.* 166 (1984), 23–43.

[56] V.I. Blagodatskikh. *Introduction to Optimal Control.* Vysshaya shkola, Moscow, 2001 (in Russian).

[57] V.I. Blagodatskikh, A.F. Filippov. Differential inclusions and optimal control. *Proc. Steklov Inst. Math..* 169 (1986), 199–259.

[58] F.S. De Blasi. Characterizations of certain classes of semicontinuous multifunctions by continuous approximations. *J. Math. Anal. Appl.* 106 (1985), no. 1, 1–18.

[59] F.S. De Blasi, L. Górniewicz, G. Pianigiani. Topological degree and periodic solutions of differential inclusions. *Nonlinear Anal. Ser. A: Theory Methods.* 37 (1999), no. 2, 217–243.

[60] F.S. De Blasi, J. Myjak. On the solutions sets for differential inclusions. *Bull. Acad. Polon. Sci. Math.* 33 (1985), no. 1–2, 17–23

[61] F.S. De Blasi, J. Myjak. On continuous approximations for multifunctions. *Pacific J. Math.* 123 (1986), no. 1, 9–31.

[62] N.A. Bobylev, S.V. Emel'yanov, S.K. Korovin. *Geometrical Methods in Variational Problems.* Mathematics and its Applications, 485. Kluwer Academic Publishers, Dordrecht, 1999.

[63] H.F. Bohnenblust, S. Karlin. On a theorem of Ville. Contribution to the theory of games, I. *Ann. Math.Studies.* 1950, Princeton, 155–160.

[64] Yu. G. Borisovich. An application of the concept of rotation of a vector field. *Soviet Math. Dokl.* 4 (1963), 1584–1587.

[65] Yu.G. Borisovich. On the relative rotation of compact vector fields in linear spaces. *Trudy Sem. Funktsion. Anal. Voronezh Gos. Univ.* 12 (1969), 3–27 (in Russian).

[66] Yu.G. Borisovich. Modern approach to the theory of topological characteristics of nonlinear operators. I. In: Global Analysis — Studies and Applications II, *Lecture Notes in Math.* 1334 (1988), 199–220.

[67] Yu.G. Borisovich. A modern approach to the theory of topological characteristics of nonlinear operators. II. In: Global Analysis — Studies and Applications IV, *Lecture Notes in Math.* 1453 (1990), 21–50.

[68] Yu.G. Borisovich, B.D. Gel'man, E. Mukhamadiev and V.V. Obukhovskii. On the rotation of multi-valued vector fields. *Soviet Math. Dokl.* 10 (1969), 956–958.

[69] Yu.G. Borisovich, B.D. Gel'man, E. Mukhamadiev, V.V. Obukhovskii. On the rotation of multivalued vector fields. *Trudy Sem. Funktsion. Anal. Voronezh Gos. Univ.* 12 (1969), 69–84 (in Russian).

[70] Yu.G. Borisovich, B.D. Gelman, A.D. Myshkis, and V.V. Obukhovskii. Topological methods in the fixed-point theory of multivalued maps. *Russian Math. Surveys.* 35 (1980), 65–143.

[71] Yu.G. Borisovich, B.D. Gelman, A.D. Myshkis, and V.V. Obukhovskii. Multivalued mapings. *J. Sov. Math.* 24 (1984), 719–791.

[72] Yu.G. Borisovich, B.D. Gelman, A.D. Myshkis, and V.V. Obukhovskii. Multivalued analysis and operator inclusions. *J. Sov. Math.* 39 (1987), 2772–2811.

[73] Yu.G. Borisovich, B.D. Gelman, A.D. Myshkis, and V.V. Obukhovskii. New results in the theory of multivalued mappings. I. Topological characteristics and solvability of operator relations. *J. Sov. Math.* 49 (1990), 800–855.

[74] Yu.G. Borisovich, B.D. Gelman, A.D. Myshkis, and V.V. Obukhovskii. *Introduction to the Theory of Multivalued Maps and Differential Inclusions.* 2nd ed., LIBROKOM, Moscow, 2011 (in Russian).

[75] Yu.G. Borisovich, Yu.E. Gliklikh. On the Lefschetz number for a class of multivalued maps. *Seventh Math. Summer School (Katsiveli 1969),* Izd. Akad. Nauk Ukrain. SSR, Kiev 1970, 283–294 (in Russian).

[76] Yu.G. Borisovich, V.V. Obukhovskii. On the optimization problem for controlled systems of parabolic type. *Proc. Stelkov Inst. Math.* 211 (1995), 85–91.

[77] K. Borsuk. *Theory of Retracts.* Monografie Matematyczne, Tom 44, Państwowe Wydawnictwo Naukowe, Warsaw, 1967.

[78] A. Bressan. On differential relations with lower continuous right-hand

side. An existence theorem. *J. Differential Equations.* 37 (1980), no. 1, 89–97.

[79] A. Bressan. On the qualitative theory of lower semicontinuous differential inclusions. *J. Differential Equations* 77 (1989), no. 2, 379–391.

[80] A. Bressan. Upper and lower semicontinuous differential inclusions: a unified approach. *Nonlinear controllability and optimal control,* Monogr. Textbooks Pure Appl. Math., 133, Dekker, New York, 1990, 21–31.

[81] A. Bressan. Differential inclusions without convexity: a survey of directionally continuous selections. *World Congress of Nonlinear Analysts '92, Vol. I–IV (Tampa, FL, 1992).* de Gruyter, Berlin, 1996, 2081–2088.

[82] A. Bressan, A. Cellina, G. Colombo. Upper semicontinuous differential inclusions without convexity. *Proc. Amer. Math. Soc.* 106 (1989), no. 3, 771–775.

[83] A. Bressan, G. Colombo. Extensions and selections of maps with decomposable values. *Studia Math.* 90 (1988), no. 1, 69–86.

[84] A. Bressan, V. Staicu. On nonconvex perturbations of maximal monotonc differential inclusions. *Set-Valued Anal.* 2 (1994), no. 3, 415–437.

[85] F.E. Browder. On a generalization of the Schauder fixed point theorem. *Duke Math. J.* 26 (1959), 291–303.

[86] F.E. Browder. The fixed point theory of multi-valued mappings in topological vector spaces. *Math. Ann.* 177 (1968), 283–301.

[87] F.E. Browder. Nonlinear Operators and Nonlinear Equations of Evolution in Banach Spaces. *Nonlinear functional analysis (Proc. Sympos. Pure Math., Vol. XVIII, Part 2, Chicago, Ill., 1968).* Amer. Math. Soc., Providence, R. I., 1976.

[88] J. Bryszewski. On a class of multi-valued vector fields in Banach spaces. *Fund. Math.* 97 (1977), no. 2, 79–94.

[89] B.M. Budak. Dispersive dynamical systems. *Vestnik Moskov. Univ.* 8 (1947), 135–137 (in Russian).

[90] B.M. Budak. The concept of motion in a generalized dynamical system. *Moskov. Gos. Univ. Uch. Zap. Mat..* 155 (1952), no. 5, 174–194 (in Russian).

[91] A.I. Bulgakov. Continuous branches of multivalued mappings, and integral inclusions with nonconvex images and their applications. I. *Differential Equations.* 28 (1992), no. 3, 303–311.

[92] A.I. Bulgakov. Continuous branches of multivalued mappings, and integral inclusions with nonconvex images and their applications. II. *Differential Equations.* 28 (1992), no. 4, 454–458.

[93] A.I. Bulgakov. Continuous branches of multivalued mappings, and integral inclusions with nonconvex images and their applications. III. *Differential Equations.* 28 (1992), no. 5, 587–592.

[94] A.I. Bulgakov, L.I. Tkach. Perturbation of a convex-valued operator by a Hammerstein-type multivalued mapping with nonconvex images, and boundary value problems for functional-differential inclusions. *Sb. Math.* 189 (1998), no. 5–6, 821–848.

[95] C. Castaing. Sur les multi-applications mesurables. *Rev. Francaise Informat. Recherche Opérationnell.* 1 (1967), no. 1, 91–126.

[96] C. Castaing. Sur les équations différentielles multivoques. *C. R. Acad. Sci. Paris Sér. A-B.* 263 (1966), A63–A66.

[97] C. Castaing, M. Valadier. *Convex Analysis and Measurable Multifunctions.* Lect. Notes in Math. 580, Springer-Verlag, Berlin–Heidelberg–New York, 1977.

[98] A. Cellina. A theorem on the approximation of compact multivalued mappings. *Atti Accad. Naz. Lincei Rend. Cl. Sci. Fis. Mat. Natur. (8).* 47 (1969), 429–433. (1970).

[99] A. Cellina. Approximation of set valued functions and fixed point theorems. *Ann. Mat. Pura Appl. (4)* 82 (1969), 17–24.

[100] A. Cellina. The role of approximation in the theory of multivalued mappings. *Differential Games and Related Topics.* (Proc. Internat. Summer School, Varenna, 1970) North-Holland, Amsterdam, 1971, 209–220.

[101] A. Cellina, A. Lasota. A new approach to the definition of topological degree for multi-valued mappings. *Atti Accad. Naz. Lincei Rend. Cl. Sci. Fis. Mat. Natur.* (8), 47 (1969), 434–440 (1970).

[102] K.C. Chang. The obstacle problem and partial differential equations with discontinuous nonlinearities. *Comm. Pure Appl. Math.* 33 (1980), no. 2, 117–146.

[103] K.C. Chang. Free boundary problems and the set-valued mappings. *J. Differential Equations* 49 (1983), no. 1, 1–28.

[104] F.H. Clarke. *Optimization and Nonsmooth Analysis.* Second edition. Classics in Applied Mathematics, 5. Society for Industrial and Applied Mathematics (SIAM), Philadelphia, PA, 1990.

[105] G. Conti, V. Obukhovskii, P. Zecca. On the topological structure of the solution set for a semilinear functional-differential inclusion in a Banach space. *Topology in nonlinear analysis (Warsaw, 1994).* Banach Center Publ., 35, Polish Acad. Sci., Warsaw, 1996, 159–169.

[106] G. Conti, V. Obukhovskii, P. Zecca. The topological degree theory

for a class of noncompact multimaps. *Ann. Mat. Pura Appl.* (4) 178 (2000), 103–113.

[107] G. Conti, J. Pejsachowicz. Fixed point theorems for multivalued weighted maps. *Ann. Mat. Pura Appl.* (4) 126 (1980), 319–341 (1981).

[108] J.-F. Couchouron, M. Kamenski. An abstract topological point of view and a general averaging principle in the theory of differential inclusions. *Nonlinear Anal. Ser. A: Theory Methods.* 42 (2000), no. 6, 1101–1129.

[109] J.-F. Couchouron, M. Kamenski, R. Precup. A nonlinear periodic averaging principle. *Nonlinear Anal.* 54 (2003), no. 8, 1439–1467.

[110] H. Covitz, S.B. Nadler, Jr. Multi-valued contraction mappings in generalized metric spaces. *Israel J. Math.* 8 (1970), 5–11.

[111] R. Cross. *Multivalued Linear Operators.* Marcel Dekker, New York, 1998.

[112] J.L. Davy. Properties of the solution set of a generalized differential equation. *Bull. Austral. Math. Soc.* 6 (1972), 379–398.

[113] G. Debreu. Integration of correspondences. *Proc. Fifth Berkeley Sympos. Math. Statist. and Probability (Berkeley, Calif., 1965/66), Vol. II: Contributions to Probability Theory, Part 1.* Univ. California Press, Berkeley, Calif., 1967, 351–372.

[114] K. Deimling. Cone-valued periodic solutions of ordinary differential equations. *Applied nonlinear analysis (Proc. Third Internat. Conf., Univ. Texas, Arlington, Tex., 1978).* Academic Press, New York-London, 1979, 127–142.

[115] K. Deimling. *Nonlinear Functional Analysis.* Springer-Verlag, Berlin-Heidelberg-New York-Tokyo, 1985.

[116] K. Deimling. *Multivalued Differential Equations.* De Gruyter Series in Nonlinear Analysis and Applications, 1. Walter de Gruyter, Berlin-New York, 1992.

[117] V.F. Dem'yanov. *A Fixed Point Theorem in Nonsmooth Analysis and Its Application.* The Publishing House of the St. Petersburg State University, St. Petersburg, 1996 (in Russian).

[118] V.F. Demyanov, A.M. Rubinov. *Constructive Nonsmooth Analysis.* Approximation end Optimization, 7. Peter Lang, Frankfurt am Main, 1995.

[119] V.F. Dem'yanov, L.V. Vasil'ev. *Nondifferentiable Optimization.* Translation Series in Mathematics and Engineering. Optimization Software, Inc., Publications Division, New York, 1985.

[120] Z. Denkowski, S. Migórski, N.S. Papageorgiou. *An Introduction to Nonlinear Analysis: Theory.* Kluwer, Boston, 2003.

[121] Z. Denkowski, S. Migórski, N.S. Papageorgiou. *An Introduction to Nonlinear Analysis: Applications.* Kluwer, Boston, 2003.

[122] Z. Ding, A.G. Kartsatos. Nonresonance problems for differential inclusions in separable Banach spaces. *Proc. Amer. Math. Soc.* 124 (1996), no. 8, 2357–2365.

[123] R. Dragoni, J.W. Macki, P. Nistri, P. Zecca. *Solution Sets of Differential Equations in Abstract Spaces.* Longman, Harlow, 1996.

[124] N. Dunford and J.T. Schwartz. *Linear Operators, Part I.* Interscience, New York, 1958.

[125] Z. Dzedzej. Fixed point index theory for a class of nonacyclic multivalued maps. *Dissertationes Math. (Rozprawy Mat.)* 253 (1985), 1–53.

[126] Z. Dzedzej, B.D. Gelman. Dimension of solution set for differential inclusions. *Demonstratio Math.* 26 (1993), 1, 149–158.

[127] S. Eilenberg, D. Montgomery. Fixed point theorems for multi-valued transformations. *Amer. J. Math.* 68 (1946), 214–222.

[128] G. Eisenack, C. Fenske. *Fixpunkttheorie.* Bibliographisches Institut, Mannheim, 1978.

[129] I. Ekeland, R. Temam. *Convex Analysis and Variational Problems.* Classics in Applied Mathematics, 28. Society for Industrial and Applied Mathematics (SIAM), Philadelphia, PA, 1999.

[130] I. Ekeland, M. Valadier. Representation of set-valued mappings. *J. Math. Anal. Appl.* 35 (1971), 621–629.

[131] R. Engelking. *General Topology.* Second edition. Sigma Series in Pure Mathematics, 6. Heldermann Verlag, Berlin, 1989.

[132] K. Fan. Fixed-point and minimax theorems in locally convex topological linear spaces. *Proc. Nat. Acad. Sci. USA.* 38 (1952), 121–126.

[133] K. Fan. A generalization of Tychonoff's fixed point theorem. *Math. Ann.* 142 (1961), no. 3, 305–310.

[134] A. Favini, A. Yagi. *Degenerate Differential Equations in Banach Spaces.* Marcel Dekker, New York, 1999.

[135] V.V. Fedorchuk, V.V. Filippov. *General Tololology. Basic Constructions.* Moscow University Press, Moscow, 1988 (in Russian).

[136] O.P. Filatov, M.M. Khapaev. *Averaging of Systems of Differential Inclusions.* Moscow University Press, Moscow, 1998 (in Russian).

[137] A.F. Filippov. On certain questions in the theory of optimal control. *Vestnik Mosk. univ. Ser. matem., mekhan. astron., phys., khim.* 2(1959), 25–32 (in Russian). Engl. transl.: *J. SIAM Control Ser. A* 1 (1962), 76–84.

[138] A.F. Filippov. Differential equations with discontinuous right-hand side. *Mat. Sb. (N.S.)* 51 (1960), no. 1, 99–128 (in Russian).

[139] A.F. Filippov. Classical solutions of differential equations with the multivalued right-hand side. *Vestnik Moskov. Univ. Ser. I Mat. Mekh.* 22 (1967) no. 3, 16–26 (in Russian).

[140] A.F. Filippov. On the existence of solutions of multivalued differential equations. *Mat. Zametki.* 10 (1971), no. 19, 307–313 (in Russian).

[141] A.F. Filippov. *Differential Equations with Discontinuous Righthand Sides.* Mathematics and its Applications (Soviet Series), 18. Kluwer Academic Publishers Group, Dordrecht, 1988.

[142] A. Fryszkowski. Continuous selections for a class of nonconvex multivalued maps. *Studia Math.* 76 (1983), no. 2, 163–174.

[143] A. Fryszkowski. *Fixed Point Theory for Decomposable Sets.* Kluwer, Dordrecht, 2004.

[144] M. Furi, M. Martelli. A degree for a class of acyclic-valued vector fields in Banach spaces. *Ann. Scuola Norm. Sup. Pisa Cl. Sci. (4) 1* (1974), 301–310 (1975).

[145] D. Gabor, W. Kryszewski. A coincidence theory involving Fredholm operators of nonnegative index. *Topol. Methods Nonlinear Anal.* 15 (2000), no.1, 43–59.

[146] G. Gabor. On the acyclicity of fixed point sets of multivalued maps. *Topol. Methods Nonlinear Anal.* 14 (1999), no. 2, 327–343.

[147] G. Gabor. Strict equilibria of multi-valued maps and common fixed points. *Z. Anal. Anwend.* 23 (2004), no. 1, 95–113.

[148] B.D. Gel'man. Multivalued integral operators and w-periodic solutions. *Voronezh Gos. Univ. Trudy Mat. Fak.* 4 (1971), 35–44 (in Russian).

[149] B.D. Gel'man. A generalization of Kakutanis fixed-point theorem to many-valued mappings. *Sov. Math. Dokl.* 14 (1973), 305–308.

[150] B.D. Gel'man. The topological characteristic of multivalued mappings and fixed-point theorems. *Sov. Math. Dokl.* 16 (1975), 360–364.

[151] B.D. Gel'man. A topological characteristic of multivalued maps in a Banach space. *Voronezh. Gos. Univ. Trudy Mat. Fak.* 16 (1975), 17–23 (in Russian).

[152] B.D. Gel'man. On the structure of a solution set for inclusions with multivalued operators. In: *Global Analysis and Mathematical Physics.* Publishing House of the Voronezh University, Voronezh 1987, 26–41 (in Russian).

[153] B.D. Gel'man. Generalized degree of multi-valued mappings. *Lect. Notes in Math.* 1520 (1992), Springer-Verlag, 173–192.

[154] B.D. Gel'man. A theorem on antipodes and coincidence points. *Russian Math. Surveys.* 51(1996), no. 1, 142–143.

[155] B.D. Gel'man. Topological properties of the set of fixed points of multivalued mappings. *Sb. Math..* 188 (1997), no. 12, 1761–1782.

[156] B.D. Gel'man. On topological dimension of a set of solutions of functional inclusions. *Differential Incl. and Optim. Control, Lect. Notes in Nonlin. Anal.*, 2 (1998), 163–178.

[157] B.D. Gel'man. Rest points of generalized dynamical systems. *Math. Notes.* 65 (1999), no. 1–2, 24–30.

[158] B.D. Gel'man. Topological dimension of the solution set of the Cauchy problem for differential inclusions. *Vestnik Voronezh. Gos. Univ. Ser. Phys. Matem.* 1 (2000), 107–115 (in Russian).

[159] B.D. Gel'man. On a class of operator equations. *Math. Notes.* 70 (2001), no 3–4, 494–501.

[160] B.D. Gel'man. A generalized implicit mapping theorem. *Funct. Anal. Appl.* 35 (2001), no. 3, 183–188.

[161] B.D. Gel'man. On the topological dimension of the solution set for operator inclusions containing surjective operators. *Vestnik Voronezh. Univ. Ser. Phys. Mat.* 1 (2001), 75–80 (in Russian).

[162] B.D. Gel'man. On the topological structure of the fixed point set for an abstract Volterra equation. *Vestnik Voronezh. Univ. Ser. Phys. Mat.* 2 (2001), 63–66 (in Russian).

[163] B.D. Gel'man. The Borsuk-Ulam theorem in infinite-dimensional Banach spaces. *Sb. Math.* 193 (2002), no. 1–2, 83–91.

[164] B.D. Gel'man. An infinite-dimensional version of the Borsuk-Ulam theorem. *Funct. Anal. Appl.* 38 (2004), no. 4, 239–242.

[165] B.D. Gel'man. Continuous approximations of multivalued mappings, and fixed points. *Math. Notes.* 78 (2005), no. 1–2, 194–203.

[166] B.D. Gelman. On the Cauchy problem for a class of degenerate differential equations with a Lipshitz right-hand side. *Funct. Anal. Appl.* 42(2008), no. 3, 227–229.

[167] B.D. Gel'man. Multivalued contractive maps and their applications. *Vestnik Voronezh. Univ. Ser. Phys. Mat.* 1 (2009), 74–86 (in Russian).

[168] B.D. Gel'man. On local solutions of degenerate differential inclusions. *Funct. Anal. Appl.* 46 (2012), no. 1, 66–68.

[169] B.D. Gel'man. The solution set of a class of equations with surjective operators. *Funct. Anal. Appl.* 49 (2015), no. 1, 60–63.

[170] B.D. Gel'man. A version of the infinite-dimensional Borsuk-Ulam

theorem for multivalued maps. *Sb. Math.* 207 (2016), no. 6, 841–853.

[171] B.D. Gel'man. How to approach nonstandard boundary value problems. *Funct. Anal. Appl.* 50 (2016), no. 1, 31–38.

[172] B.D. Gel'man. Periodic trajectories and coincidence points of tuples of set-valued maps. *Funct. Anal. Appl.* 52 (2018), no. 2, 139–143.

[173] B.D. Gel'man. The Caristi inequality and α-contractive mappings. *Funct. Anal. Appl.* 53 (2019), no. 3, 224–228.

[174] B.D. Gel'man, V.V. Obukhovskii. Some fixed point theorems for multivalued maps of condensing type. In: *Algebraic Problems of Analysis and Topology* Voronezh Univ. Press, Voronezh 1990, 110–115 (in Russian).

[175] B.D. Gelman, V.V. Obukhovskii. New results in the theory of multivalued mappings. II. Analysis and applications. *J. Sov. Math.* 64 (1993), 853–883.

[176] B. Gel'man, V. Obukhovskii. On connections between some selection and approximation results for multivalued maps. *Topology Appl.* 155 (2008), no. 8, 805–813.

[177] B.D. Gel'man, V.V. Obukhovskii. On the fixed points of multivalued maps of acyclic type. J. Math. Sci. (N.Y.) 225 (2017), no. 4, 565–574.

[178] B.D. Gel'man, V. Obukhovskii, H.R. Al-Hashemi. On some approximation methods in the theory of operator inclusions. *Fixed Point Theory* 7 (2006), no. 2, 245–261.

[179] E. Getmanova, V. Obukhovskii. A note on random equilibrium points of two multivalued maps. *J. Nonlin. Variat. Anal.* 2 (2018), no. 3, 269–272.

[180] E. Getmanova, V. Obukhovskii, J.-C. Yao. A random topological fixed point index for a class of multivalued maps. *Appl. Set-Valued Anal. Optim.* 1 (2019), no. 2, 95–103.

[181] K. Glashoff, J. Sprekels. An application of Glicksberg's theorem to set-valued integral equations arising in the theory of thermostats. *SIAM J. Math. Anal.* 12 (1981), no. 3, 477–486.

[182] I.L. Glicksberg. A further generalization of the Kakutani fixed point theorem, with application to Nash equilibrium points. *Proc. Amer. Math. Soc.* 3 (1952), 1, 170–174.

[183] Yu.E. Gliklikh, A.V. Obukhovskii. On a two-point boundary value problem for second-order differential inclusions on Riemannian manifolds. *Abstr. Appl. Anal.* 2003, no. 10, 591–600.

[184] B. Glodde, H.-D. Niepage. *Einführung in die Mengenwertige Analysis und die Theorie der Kontingentgleichungen.* Seminarberichte, 41. Humboldt Universität, Sektion Mathematik, Berlin, 1981.

[185] K. Goebel, W.A. Kirk. *Topics in Metric Fixed Point Theory.* Cambridge Univ. Press, Cambridge, 1990.

[186] V.V. Goncharov, A.A. Tolstonogov. Common continuous selections of multivalued mappings with nonconvex values, and their applications. *Math. USSR-Sb.* 73 (1992), no.2, 319–339.

[187] C. Gori, V. Obukhovskii, M. Ragni, P. Rubbioni. Existence and continuous dependence results for semilinear functional differential inclusions with infinite delay. *Nonlinear Anal. Ser. A: Theory Methods.* 51 (2002), no. 5, 765–782.

[188] C. Gori, V. Obukhovskii, M. Ragni, P. Rubbioni. On some properties of semilinear functional differential inclusions in abstract spaces. *J. Concr. Appl. Math.* 4 (2006), no. 2, 183–214.

[189] C. Gori, V. Obukhovskii, P. Rubbioni, V. Zvyagin. Optimization of the motion of a visco-elastic fluid via multivalued topological degree method. *Dynam. Systems Appl.* 16 (2007), no. 1, 89–104.

[190] L. Górniewicz. *Topological Fixed Point Theory of Multivalued Mappings.* Second edition. Topological Fixed Point Theory and Its Applications, 4. Springer, Dordrecht, 2006.

[191] L. Górniewicz, A. Granas, W. Kryszewski. On the homotopy method in the fixed point index theory of multi-valued mappings of compact absolute neighborhood retracts. *J. Math. Anal. Appl.* 161 (1991), no. 2, 457–473.

[192] L. Górniewicz, Z. Kucharski. Coincidence of k-set contraction pairs. *J. Math. Anal. Appl.* 107 (1985), no. 1, 1–15.

[193] L. Górniewicz, P. Nistri, V. Obukhovskii. Differential inclusions on proximate retracts of Hilbert spaces. *Internat. J. Non-Linear Differential Equat. Theory, Methods, Appl.* 3 (1997), 13–26.

[194] L. Górniewicz, S. Plaskacz. Periodic solutions of differential inclusions in R^n. *Boll. Un. Mat. Ital.* A (7), 7 (1993), no. 3, 409–420.

[195] S. Graf. Selected results on measurable selections. *Rend. Circ. Mat. Palermo.* (2) 1982, 87–122.

[196] A. Granas. Sur la notion du degré topologique pour une certaine classe de transformations multivalentes dans les espaces de Banach. *Bull. Acad. Polon. Sci. Sér. Sci. Math. Astr. Phys.* 7 (1959), no. 4, 191–194.

[197] A. Granas. Theorem on antipodes and theorems on fixed points for a

certain class of multi-valued mappings in Banach spaces. *Bull. Acad. Polon. Sci. Sér. Sci. Math. Astr. Phys.* 7 (1959), no. 5, 271–275.

[198] A. Granas, J. Dugundji. *Fixed Point Theory.* Springer-Verlag, New York, 2003.

[199] A. Granas, J.W. Jaworowski. Some theorems on multi-valued mappings of subsets of the Euclidean space. *Bull. Acad. Polon. Sci. Sér. sci. math., astron. et phys.* 7 (1959), no. 5, 277–283.

[200] I. Gubanova, V. Obukhovskii, C.-F. Wen. On some optimization problems for a class of fractional order feedback control systems. *Appl. Anal. Optim.* 2 (2018), no. 1, 47–57.

[201] J. Gwinner. On fixed points and variational inequalities—a circular tour. *Nonlinear Anal.* 5 (1981), no. 5, 565–583.

[202] G. Haddad, J.-L. Lasry. Periodic solutions of functional-differential inclusions and fixed points of σ-selectionable correspondences. *J. Math. Anal. Appl.* 96 (1983), no. 2, 295–312.

[203] O. Hadžić. *Fixed Point Theory in Topological Vector Spaces.* Univerzitet u Novom Sadu, Institut za Matematiku, Novi Sad, 1984.

[204] Ph. Hartman. *Ordinary Differential Equations.* Corrected reprint of the second (1982) edition. Classics in Applied Mathematics, 38. Society for Industrial and Applied Mathematics (SIAM), Philadelphia, PA, 2002.

[205] F. Hiai, H. Umegaki. Integrals, conditional expectations, and martingales of multivalued functions. *J. Multivariate Anal.* 7 (1977), no. 1, 149–182.

[206] W. Hildenbrand. *Core and Equilibria of a Large Economy.* Princeton Studies in Mathematical Economics, 5. Princeton University Press, Princeton, N.J., 1974.

[207] C.J. Himmelberg. Measurable relations. *Fund. Math.* 87 (1975), 53–72.

[208] C.J. Himmelberg, M.Q. Jacobs, F.S. Van Vleck. Measurable multifunctions, selectors, and Filippov's implicit functions lemma. *J. Math. Anal. Appl.* 25 (1969), 276–284.

[209] C.J. Himmelberg, F.S. Van Vleck. A note on the solution sets of differential inclusions. *Rocky Mountain J. Math.* 12 (1982), no. 4, 621–625.

[210] S. Hu, N.S. Papageorgiou. *Handbook of Multivalued Analysis. Vol. I. Theory.* Kluwer, Dordrecht, 1997.

[211] S. Hu, N.S. Papageorgiou. *Time-Dependent Subdifferential Evolution Inclusions and Optimal Control.* Mem. Amer. Math. Soc. 133 (1998), no. 632.

[212] S. Hu, N.S. Papageorgiou. *Handbook of Multivalued Analysis. Vol. II. Applications.* Kluwer, Dordrecht, 2000.

[213] M. Hukuhara. Sur l'application semi-continue dont la valeur est un compact convexe. *Funkcial. Ekvac.* 10 (1967), 43–66.

[214] M. Hukuhara. Intégration des applications mesurables dont la valeur est un compact convexe. *Funkcial. Ekvac.* 10 (1967), 205–223.

[215] A.D. Ioffe, V.L. Levin. Subdifferentials of convex functions. *Trudy Moskov. Mat. Ob.* 26 (1972), 3–73 (in Russian).

[216] A.D. Ioffe, V.M. Tihomirov. *Theory of Extremal Problems.* Studies in Mathematics and its Applications, 6. North-Holland Publishing Co., Amsterdam-New York, 1979.

[217] Ya.A. Izrailevich, V.V. Obukhovskii. On equivariant multi-valued maps. *Soviet Math. Dokl.* 13 (1972), 864–867.

[218] M.Q. Jacobs. Remarks on some recent extensions of Filippov's implicit functions lemma. *SIAM J. Control.* 5 (1967), 622–627.

[219] M.Q. Jacobs. Measurable multivalued mappings and Lusin's theorem. *Trans. Amer. Math. Soc.* 134 (1968), 471–481.

[220] S. Kakutani. A generalization of Brouwer's fixed point theorem. *Duke Math. J.* 8 (1941), 457–459.

[221] M. Kamenskii, P. Nistri. An averaging method for singularly perturbed systems of semilinear differential inclusions with C_0-semigroups. *Set-Valued Anal.* 11 (2003), no. 4, 345–357.

[222] M.I. Kamenskii, P. Nistri, V.V. Obukhovskii, P. Zecca. Optimal feedback control for a semilinear evolution equation. *J. Optim. Theory Appl.* 82 (1994), no. 3, 503–517.

[223] M.I. Kamenskii, P. Nistri, P. Zecca. On the periodic solutions problem for parabolic inclusions with a large parameter. *Topol. Methods Nonlinear Anal.* 8 (1996), no. 1, 57–77.

[224] M.I. Kamenskii, V.V. Obukhovskii. On periodic solutions of differential inclusions with unbounded operators in Banach spaces. *Zb. Rad. Prirod.-Mat. Fak. Ser. Mat.* 21 (1991), no. 1, 173–191.

[225] M.I. Kamenskii, V.V. Obukhovskii. Condensing multioperators and periodic solutions of parabolic functional-differential inclusions in Banach spaces. *Nonlinear Anal.* 20 (1993), no. 7, 781–792.

[226] M.I. Kamenskii, V.V. Obukhovskii. The operator of translation along trajectories of semilinear controlled systems. *Differential Equations.* 32 (1996), 755–762.

[227] M. Kamenskii, V. Obukhovskii, G. Petrosyan, J.-C. Yao. On semilinear fractional order differential inclusions in Banach spaces. *Fixed Point Theory* 18 (2017), no. 1, 269–291.

[228] M. Kamenskii, V. Obukhovskii, G. Petrosyan, J.-C. Yao. On approximate solutions for a class of semilinear fractional-order differential equations in Banach spaces. *Fixed Point Theory Appl.* 2017, Paper No. 28, 20 pp.

[229] M. Kamenskii, V. Obukhovskii, G. Petrosyan, J.-C. Yao. Boundary value problems for semilinear differential inclusions of fractional order in a Banach space. *Appl. Anal.* 97 (2018), no. 4, 571–591.

[230] M. Kamenskii, V. Obukhovskii, G. Petrosyan, J.-C. Yao. Existence and approximation of solutions to nonlocal boundary value problems for fractional differential inclusions. *Fixed Point Theory Appl.* 2019, Paper No. 2, 21 pp.

[231] M. Kamenskii, V. Obukhovskii, G. Petrosyan, J.-C. Yao. On a periodic boundary value problem for a fractional-order semilinear functional differential inclusions in a Banach space. *Mathematics* 7 (2019), no. 12, 5–19.

[232] M. Kamenskii, V. Obukhovskii, J.-C. Yao. On some topological methods in theory of neutral type operator differential inclusions with applications to control systems. *Discuss. Math. Differ. Incl. Control Optim.* 33 (2013), no. 2, 193–204.

[233] M. Kamenski, V. Obukhovski, P. Zecca. On the translation multi-operator along the solutions of semilinear differential inclusions in Banach spaces. *Canad. Appl. Math. Quart.* 6 (1998), no. 2, 139–155.

[234] M. Kamenskii, V. Obukhovskii, P. Zecca. On semilinear differential inclusions with lower semicontinuous nonlinearities. *Ann. Mat. Pura Appl.* (4), 178 (2000), 235–244.

[235] M. Kamenskii, V. Obukhovskii, P. Zecca. *Condensing Multivalued Maps and Semilinear Differential Inclusions in Banach Spaces.* De Gruyter Series in Nonlinear Analysis and Applications, 7. Walter de Gruyter, Berlin-New York, 2001.

[236] S. Karlin. *Mathematical Methods and Theory in Games, Programming, and Economics.* Reprint of the 1959 original. Dover Publications, Inc., New York, 1992.

[237] T.D. Ke, N.V. Loi, V. Obukhovskii. Decay solutions for a class of fractional differential variational inequalities. *Fract. Calc. Appl. Anal.* 18 (2015), no. 3, 531–553.

[238] T.D. Ke, V. Obukhovskii. Controllability for systems governed by second-order differential inclusions with nonlocal conditions. *Topol. Methods Nonlinear Anal.* 42 (2013), no. 2, 377–403.

[239] T.D. Ke, V. Obukhovskii, N.-C. Wong, J.-C. Yao. An abstract

Cauchy problem for higher order functional differential inclusions with infinite delay. *Discuss. Math. Differ. Incl. Control Optim.* 31 (2011), no. 2, 199–229.

[240] T.D. Ke, V. Obukhovskii, N.-C. Wong, J.-C. Yao. On a class of fractional order differential inclusions with infinite delays. *Appl. Anal.* 92 (2013), no. 1, 115–137.

[241] J.L. Kelley. *General Topology.* Reprint of the 1955 edition. Graduate Texts in Mathematics, 27. Springer-Verlag, New York-Berlin, 1975.

[242] N. Kikuchi. Control problems of contingent equation. *Publ. Res. Inst. Math. Sci. Ser. A* 3 (1967/1968), 85–99.

[243] N. Kikuchi. On some fundamental theorems of contingent equations in connection with the control problems. *Publ. Res. Inst. Math. Sci. Ser. A* 3 (1967/1968), 177–201.

[244] N. Kikuchi. On contingent equations satisfying the Carathéodory type conditions. *Publ. Res. Inst. Math. Sci. Ser. A* 3 (1967/1968), 361–371.

[245] M. Kisielewicz. *Differential Inclusions and Optimal Control.* Kluwer, Dordrecht; PWN-Polish Scientific Publishers, Warsaw, 1991.

[246] E. Klein, A.C. Thompson. *Theory of Correspondences. Including Applications to Mathematical Economics.* Wiley, New York, 1984.

[247] A.N. Kolmogorov, S.V. Fomin. *Elements of the Theory of Functions and Functional Analysis.* Dover Publications, Inc., Mineola, NY, 1999.

[248] S.V. Kornev. On the method of multivalent guiding functions in the problem of periodic solutions of differential inclusions. Autom. Remote Control 64 (2003), no. 3, 409–419.

[249] S.V. Kornev. Nonsmooth integral directing functions in the problems of forced oscillations. Autom. Remote Control 76 (2015), no. 9, 1541–1550.

[250] S.V. Kornev. Multivalent guiding functions in a problem on the existence of periodic solutions of some classes of differential inclusions. Russian Math. (Iz. VUZ) 60 (2016), no. 11, 11–21.

[251] S. Kornev, N.V. Loi, V. Obukhovskii, C.-F. Wen. Random nonsmooth integral guiding functions and asymptotic behavior of trajectories for random differential inclusions. *J. Nonlinear Convex Anal.* 19 (2018), no. 3, 493–500.

[252] S.V. Kornev, V.V. Obukhovskii. On nonsmooth multivalent guiding functions. *Differential Equations.* 39(2003), no. 11, 1578–1584.

[253] S.V. Kornev, V.V. Obukhovskii. On some versions of the topological

degree theory for nonconvex-valued multimaps. *Voronezh. Gos. Univ. Trudy Mat. Fak. (N.S.)* 8 (2004), 56–74 (in Russian).

[254] S. Kornev, V. Obukhovskii. On some developments of the method of integral guiding functions. *Functional Differential Equat.* 12 (2005), no. 3–4, 303–310.

[255] S.V. Kornev, V.V. Obukhovskii. Nonsmooth guiding potentials in problems of forced oscillations. *Autom. Remote Control* 68 (2007), no. 1, 1–8.

[256] S.V. Kornev, V.V. Obukhovskii. Localization of the method of guiding functions in the problem of periodic solutions of differential inclusions. *Russian Math. (Iz. VUZ)* 53 (2009), no. 5, 19–27.

[257] S.V. Kornev, V.V. Obukhovskii. Asymptotic behavior of solutions of differential inclusions and the method of guiding functions. *Differ. Equ.* 51 (2015), no. 6, 711–716.

[258] S. Kornev, V. Obukhovskii, P. Zecca. Guiding functions and periodic solutions for inclusions with causal multioperators. *Appl. Anal.* 96 (2017), no. 3, 418–428.

[259] S. Kornev, V. Obukhovskii, P. Zecca. On multivalent guiding functions method in the periodic problem for random differential equations. *J. Dynam. Differential Equations.* 31 (2019), no. 2, 1017–1028.

[260] M.A. Krasnoselskii. *The Operator of Translation along the Trajectories of Differential Equations.* Amer. Math. Soc., Providence, RI., 1968.

[261] M.A. Krasnoselskii, A.V. Pokrovskii. *Systems with Hysteresis.* Springer-Verlag, Berlin-Heidelberg-New York, 1989.

[262] M.A. Krasnoselskii, P.P. Zabreiko. *Geometrical Methods of Nonlinear Analysis.* Springer-Verlag, Berlin, 1984.

[263] N.N. Krasovskii, A.I. Subbotin. *Positional Differential Games.* Nauka, Moscow, 1974 (in Russian).

[264] W. Kryszewski. Topological and approximation methods of degree theory of set-valued maps. *Dissertationes Math. (Rozprawy Mat.)* 336 (1994), 1–101.

[265] W. Kryszewski. *Homotopy Properties of Set-Valued Mappings.* Univ. N. Copernicus Publishing, Toruń, 1997.

[266] W. Kryszewski. Topological structure of solution sets of differential inclusions: the constrained case. *Abstr. Appl. Anal.* 2003, no. 6, 325–351.

[267] Z. Kucharski. A coincidence index. *Bull. Acad. Polon. Sci. Sér. Sci. Math. Astronom. Phys.* 24 (1976), no. 4, 245–252.

[268] Z. Kucharski. Two consequences of the coincidence index. *Bull. Acad. Polon. Sci. Sér. Sci. Math. Astronom. Phys.* 24 (1976), no. 6, 437–444.

[269] K. Kuratowski. Les fonctions semi-continues dans l'éspace des ensembles fermés. *Fund. Math.* 18 (1931), 148–159.

[270] K. Kuratowski. *Topology, Vols I and II,* 4th. ed., Panstowowe Wyd Nauk., Warszawa (Academic Press, New York, 1966).

[271] K. Kuratowski, C. Ryll-Nardzewski. A general theorem on selectors. *Bull. Acad. Polon. Sci. Sér. Sci. Math. Astronom. Phys.* 13 (1965), 397–403.

[272] A.V. Kurzhanskii. *Control and Observation under Uncertainty.* Nauka, Moscow, 1977 (in Russian).

[273] K. Lancaster. *Mathematical Economics.* Macmillan, New York, 1968.

[274] A. Lasota, Z. Opial. An application of the Kakutani–Ky Fan theorem in the theory of ordinary differential equations. *Bull. Acad. Polon. Sci. Sér. Sci. Math. Astronom. Phys.* 13 (1965), 11–12, 781–786.

[275] A. Lasota, Z. Opial. Fixed-point theorems for multi-valued mappings and optimal control problems. *Bull. Acad. Polon. Sci. Sér. Sci. Math. Astronom. Phys.* 16 (1968), 8, 645–649.

[276] A. Lasota, Z. Opial. An approximation theorem for multi-valued mappings. *Podstawy Sterowania* 1 (1971), 1, 71–75.

[277] J.-M. Lasry, R. Robert. Degré topologique pour certains couples de fonctions et applications aux équations diffe'rentielles multivoques. *C. R. Acad. Sci. Paris Sér. A-B* 283 (1976), no. 4, Aii, A163–A166.

[278] J.-M. Lasry, R. Robert. Acyclicité de l'ensemble des solutions de certaines équations fonctionnelles. C. R. Acad. Sci. Paris Sér. A-B 282 (1976), no. 22, Aii, A1283–A1286.

[279] J.-M. Lasry, R. Robert. *Analyse non Lineaire Multivoque.* Publ. no. 7611. Centre de Recherche de Mathem. de la Decis. Ceremade, Univ. de Paris, Dauphine, 1977.

[280] Y.-C. Liou, V. Obukhovskii, J.-C. Yao. Application of a coincidence index to some classes of impulsive control systems. *Nonlinear Anal.* 69 (2008), no. 12, 4392–4411.

[281] Y.-C. Liou, V. Obukhovskii, J.-C. Yao. Controllability for a class of degenerate functional differential inclusions in a Banach space. *Taiwanese J. Math.* 12 (2008), no. 8, 2179–2200.

[282] Y.-C. Liou, V. Obukhovskii, J.-C. Yao. On topological index of solutions for variational inequalities on Riemannian manifolds. *Set-Valued Var. Anal.* 20 (2012), no. 3, 369–386.

[283] Z. Liu, N.V. Loi, V. Obukhovskii. Existence and global bifurcation of periodic solutions to a class of differential variational inequalities. *Internat. J. Bifur. Chaos Appl. Sci. Engrg.* 23 (2013), no. 7, 1350125, 10 pp.

[284] N.V. Loi, T.D. Ke, V. Obukhovskii, P. Zecca. Topological methods for some classes of differential variational inequalities. *J. Nonlinear Convex Anal.* 17 (2016), no. 3, 403–419.

[285] N.V. Loi, T.D. Ke, M.Q. Vu, V. Obukhovskii. Random integral guiding functions with application to random differential complementarity systems. *Discuss. Math. Differ. Incl. Control Optim.* 38 (2018), no. 1–2, 113–132.

[286] N.V. Loi, V. Obukhovskii. On the global bifurcation for solutions of linear Fredholm inclusions with convex-valued perturbations. *Fixed Point Theory* 10 (2009), no. 2, 289–303.

[287] N.V. Loi, V. Obukhovskii. On global bifurcation of periodic solutions for functional differential inclusions. *Functional Differential Equat.* 17 (2010), no.1–2, 161–172.

[288] N.V. Loi, V. Obukhovskii. On the existence of solutions for a class of second-order differential inclusions and applications. *J. Math. Anal. Appl.* 385 (2012), no. 1, 517–533.

[289] N.V. Loi, V. Obukhovskii. Guiding functions for generalized periodic problems and applications. *Appl. Math. Comput.* 218 (2012), no. 24, 11719–11726.

[290] N.V. Loi, V. Obukhovskii, J.-C. Yao. A bifurcation of solutions of nonlinear Fredholm inclusions involving CJ-multimaps with applications to feedback control systems. *Set-Valued Var. Anal.* 21 (2013), no. 2, 247–269.

[291] N.V. Loi, V. Obukhovskii, J.-C. Yao. A multiparameter global bifurcation theorem with application to a feedback control system. *Fixed Point Theory.* 16 (2015), no. 2, 353–370.

[292] N.V. Loi, V. Obukhovskii, P. Zecca. Non-smooth guiding functions and periodic solutions of functional differential inclusions with infinite delay in Hilbert spaces. *Fixed Point Theory.* 13 (2012), no. 2, 565–582.

[293] N.V. Loi, V. Obukhovskii, P. Zecca. On the global bifurcation of periodic solutions of differential inclusions in Hilbert spaces. *Nonlinear Anal.* 76 (2013), 80–92.

[294] N.V. Loi, M.Q. Vu, N.T. Hoai, V. Obukhovskii. A nonlocal problem for projected differential equations and inclusions with applications. *Fixed Point Theory.* 20 (2019), no. 1, 233–244.

[295] S. Lojasiewicz, Jr. The existence of solutions for lower semicontinuous orientor fields. *Bull. Acad. Polon. Sci. Sér. Sci. Math.* 28 (1980), no. 9–10, 483–487 (1981).

[296] L. Lu, Z. Liu, V. Obukhovskii. Second order differential variational inequalities involving anti-periodic boundary value conditions. *J. Math. Anal. Appl.* 473 (2019), no. 2, 846–865.

[297] T.-W. Ma. Topological degrees of set-valued compact fields in locally convex spaces. *Dissertationes Math. (Rozprawy Mat.)* 92 (1972), 1–43.

[298] J.W. Macki, P. Nistri, P. Zecca. Measurable and directionally continuous selections for the control of uncertain systems. *Dynam. Contin. Discrete Impuls. Systems.* 2 (1996), no. 4, 397–409.

[299] V.L. Makarov, A.M. Rubinov. *Mathematical Theory of Economic Dynamics and Equilibria.* Springer-Verlag, New York-Heidelberg, 1977.

[300] L. Malaguti. Monotone trajectories of differential inclusions in Banach spaces. *J. Convex Anal.* 3 (1996), no. 2, 269–281.

[301] A. Marchaud. Sur les champs continus de demi-cônes convexes et leurs intégrales. *C. R. Acad. Sci. Paris* 199 (1934), n. 23, 1278–1280.

[302] A. Marchaud. Sur les champs de demi-cônes et equations differentielles du premier ordre. *Bull. Soc. Math. France.* 62 (1934), 1–38.

[303] A. Marchaud. Sur les champs continus de demi-cônes convexes et leurs integrales. *Comp. Math.* 3 (1936), n. 1, 89–127.

[304] A. Marchaud. Sur les champs de demi-cônes convexes. *Bull. Sci. Math.* 62 (1938), 229–240.

[305] J. T. Markin. A fixed point theorem for set valued mappings. *Bull. Amer. Math. Soc.* 74 (1968), 639–640.

[306] E. Michael. Continuous selections, I. *Ann. Math.* 63 (1956), 2, 361–381.

[307] E. Michael. A survey of continuous selections. *Lect. Notes in Math.* 171, Springer-Verlag, Berlin, 1970, 54–58.

[308] E. Michael. Continuous selections avoiding a set. *Topology Appl.* 28 (1988), no. 3, 195–213.

[309] M.D.P. Monteiro Marques. *Differential Inclusions in Nonsmooth Mechanical Problems. Shocks and Dry Friction.* Birkhäuser Verlag, Basel, 1993.

[310] B.S. Mordukhovich. *Approximation Methods in Problems of Optimization and Control.* John Wiley end Sons, Inc., New York, 1998.

[311] B.S. Mordukhovich. *Variational Analysis and Generalized Differentiation. I. Basic Theory.* Springer, Berlin, 2006.

[312] B.S. Mordukhovich. *Variational Analysis and Generalized Differentiation. II. Applications.* Springer, Berlin, 2006.

[313] A.D. Myshkis. General theory of differential equations with retarded argument. *Uspehi Matem. Nauk (N.S)* 4 (1949), no. 5(33), 99–141 (in Russian).

[314] A.D. Myshkis. Generalizations of the theorem on a stationary point of a dynamical system inside a closed trajectory. *Mat. Sb.* 34 (1954), no. 3, 525-540 (in Russian).

[315] S.B. Nadler, Jr. Multi-valued contraction mappings. *Pacific J. Math.* 30 (1969), 475–488.

[316] I.P. Natanson. *Theory of Functions of a Real Variable.* Dover Publications, Inc., Mineola, NY, 2016.

[317] V.V. Nemytskii The fixed point method in analysis. *Uspekhi Mat. Nauk* 1(1936), 141–174 (in Russian).

[318] J. von Neumann. Über ein ökonomisches Gleichungssystem und eine Verallgemeinerung des Brouwerschen Fixpunktsaltzes. *Ergebnisse eines Math. Kolloquiums, 8. 1935-36.* Leipzig-Wien, 1937, 73–83.

[319] H. Nikaido. *Convex Structures and Economic Theory.* Academic Press, New York, 1968.

[320] P. Nistri, V.V. Obukhovskii, P. Zecca. On the solvability of systems of inclusions involving noncompact operators. *Trans. Amer. Math. Soc.* 342 (1994), no. 2, 543–562.

[321] P. Nistri, V. Obukhovskii, P. Zecca. Viability for feedback control systems in Banach spaces via Carathéodory closed-loop controls. *Differential Equations Dynam. Systems* 4 (1996), no. 3–4, 367–378.

[322] V.V. Obukhovskii. On some fixed point principles for multivalued condensing operators. *Trudy Mat. Fac. Voronezh Univ.* 4 (1971), 70–79 (in Russian).

[323] V.V. Obukhovskii. Periodic solutions of control systems. *Trudy Mat. Fac. Voronezh Univ.* 7 (1972), 68–76 (in Russian).

[324] V.V. Obukhovskii. On periodic solutions of differential equations with multi–valued right-hand side. *Trudy Mat. Fac. Voronezh Univ.* 10 (1973), 74–82 (in Russian).

[325] V.V. Obukhovskii. Asymptotic fixed pont theorems and stationary points of dynamical systems without uniqueness. *Trudy aspirantov Voronezh Gos. Univ.* 2 (1974), 30–38 (in Russian).

[326] V.V. Obukhovskii. A condition for positive equilibrium in a model of a competitive economy. *Proceedings of the Sixth Winter School*

on Mathematical Programming and Related Questions (Drogobych, 1973). Central. Ekonom.-Mat. Inst. Akad. Nauk SSSR, Moscow, 1975, 183–189 (in Russian).

[327] V.V. Obukhovskii. On the topological degree for a class of noncompact multivalued mappings. *Funktsionalnyi Analiz (Ulyanovsk).* 23 (1984), 82–93 (in Russian).

[328] V.V. Obukhovskii. Semilinear functional-differential inclusions in a Banach space and controlled parabolic systems. *Soviet J. Automat. Inform. Sci.* 24 (1991), 71–79.

[329] V. Obukhovskii. Topological methods in some optimization problems for systems governed by differential inclusions. *Optimization.* 60 (2011), no. 6, 671–683.

[330] V. Obukhovskii, N.V. Loi, S. Kornev. Existence and global bifurcation of solutions for a class of operator-differential inclusions. *Differ. Equ. Dyn. Syst.* 20 (2012), no. 3, 285–300.

[331] V. Obukhovskii, P. Rubbioni. On a controllability problem for systems governed by semilinear functional differential inclusions in Banach spaces. *Topol. Methods Nonlinear Anal.* 15 (2000), no. 1, 141–151.

[332] V. Obukhovskii, T. Starova. On stability of fixed points of multivalued maps. *Topol. Methods Nonlinear Anal.* 17 (2001), no. 1, 133–141.

[333] V. Obukhovskii, P. Zecca. On some properties of dissipative functional differential inclusions in a Banach space. *Topol. Methods Nonlinear Anal.* 15 (2000), no. 2, 369–384.

[334] V. Obukhovskii, P. Zecca. A Kneser type property for the solution set of a semilinear differential inclusion with lower semicontinuous nonlinearity. *Set valued mappings with applications in nonlinear analysis.* Ser. Math. Anal. Appl., 4, Taylor and Francis, London, 2002, 369–381.

[335] V. Obukhovskii, P. Zecca. On boundary value problems for degenerate differential inclusions in Banach spaces. *Abstr. Appl. Anal.* 2003, no. 13, 769–784.

[336] V. Obukhovski, P. Zecca. Controllability for systems governed by semilinear differential inclusions in a Banach space with a noncompact semigroup. *Nonlinear Anal.* 70 (2009), no. 9, 3424–3436.

[337] V. Obukhovskii, P. Zecca. On semilinear differential inclusions in Banach spaces with nondensely defined operators. *J. Fixed Point Theory Appl.* 9 (2011), no. 1, 85–100.

[338] V. Obukhovskii, P. Zecca. On certain classes of functional inclusions

with causal operators in Banach spaces. *Nonlinear Anal.* 74 (2011), no. 8, 2765–2777.

[339] V. Obukhovskii, P. Zecca, N.V. Loi, S. Kornev. *Method of Guiding Functions in Problems of Nonlinear Analysis.* Lect. Notes in Math. 2076, Springer-Verlag, Berlin-Heidelberg, 2013.

[340] V. Obukhovskii, P. Zecca, V. Zvyagin. On coincidence index for multivalued perturbations of nonlinear Fredholm maps and some applications. *Abstr. Appl. Anal.* 7 (2002), no. 6, 295–322.

[341] V. Obukhovskii, P. Zecca, V.Zvyagin. Optimal feedback control in the problem of the motion of a viscoelastic fluid. *Topol. Methods Nonlinear Anal.* 23 (2004), no. 2, 323–337.

[342] V. Obukhovskii, P. Zecca, V. Zvyagin. An oriented coincidence index for nonlinear Fredholm inclusions with nonconvex-valued perturbations. *Abstr. Appl. Anal.* 2006, Art. ID 51794, 21 pp.

[343] V. Obukhovskii, P. Zecca, V. Zvyagin. On some generalizations of the Landesman-Lazer theorem. *Fixed Point Theory* 8 (2007), no. 1, 69–85.

[344] C. Olech. A note concerning set-valued measurable functions. *Bull. Acad. Polon. Sci. Sér. Sci. Math. Astronom. Phys.* 13 (1965), 317–321.

[345] C. Olech. Existence of solutions of non-convex orientor fields. *Boll. Un. Mat. Ital. (4).* 11 (1975), no. 3, suppl., 189–197.

[346] V.I. Opoitsev. *Equilibrium and Stability in Models of Collective Behavior.* Nauka, Moscow, 1977 (in Russian).

[347] N.S. Papageorgiou. On multivalued evolution equations and differential inclusions in Banach spaces. *Comment. Math. Univ. St. Paul.* 36 (1987), no. 1, 21–39.

[348] N.S. Papageorgiou. Boundary value problems for evolution inclusions. *Comment. Math. Univ. Carolin.* 29 (1988), no. 2, 355–363.

[349] N.S. Papageorgiou. On multivalued semilinear evolution equations. *Boll. Un. Mat. Ital.* B (7), 3 (1989), no. 1, 1–16.

[350] N.S. Papageorgiou. Boundary value problems and periodic solutions for semilinear evolution inclusions. *Comment. Math. Univ. Carolin.* 35 (1994), no. 2, 325–336.

[351] T. Parthasarathy. *Selection Theorems and Their Applications.* Lecture Notes in Mathematics, 263. Springer-Verlag, Berlin-New York, 1972.

[352] N.A. Perestyuk, V.A. Plotnikov, A.M. Samoilenko, N.V. Skripnik. *Differential Equations with Impulse Effects. Multivalued Right-Hand*

Sides with Discontinuities. De Gruyter Studies in Mathematics, 40. Walter de Gruyter, Berlin, 2011.

[353] A. Petruşel, G. Moţ. *Multivalued Analysis and Mathematical Economics*. House of the Book of Science, Cluj-Napoca, 2004.

[354] W.V. Petryshyn, P.M. Fitzpatrick. A degree theory, fixed point theorems, and mapping theorems for multivalued noncompact mappings. *Trans. Amer. Math. Soc.* 194 (1974), 1–25.

[355] S. Plaskacz. Periodic solutions of differential inclusions on compact subsets of R^n. *J. Math. Anal. Appl.* 148 (1990), no. 1, 202–212.

[356] S. Plaskacz. On the solution sets for differential inclusions. *Boll. Un. Mat. Ital.* A (7), 6 (1992), no. 3, 387–394.

[357] A. Pliś. Remark on measurable set-valued functions. *Bull. Acad. Polon. Sci. Sér. Sci. Math. Astronom. Phys.* 9 (1961), 857–859.

[358] A. Pliś. Trajectories and quasitrajectories of an orientor field. *Bull. Acad. Polon. Sci. Sér. Sci. Math. Astronom. Phys.* 11 (1963) 369–370.

[359] A. Pliś. Measurable orientor fields. *Bull. Acad. Polon. Sci. Sér. Sci. Math. Astronom. Phys.* 13 (1965), 565–569.

[360] A. Pliś. On trajectories of orientor fields. *Bull. Acad. Polon. Sci. Sér. Sci. Math. Astronom. Phys.* 13 (1965), 571–573.

[361] V.A. Plotnikov, A.V. Plotnikov, A.N. Vityuk. *Differential Equations with a Multivalued Right-Hand Side. Asymptotic Methods*. Astro-Print, Odessa, 1999 (in Russian).

[362] E.S. Polovinkin. *Set-Valued Analysis and Differential Inclusions*. Fizmatlit, Moscow, 2014 (in Russian).

[363] E.S. Polovinkin, M.V. Balashov. *Elements of Convex and Strongly Convex Analysis*. Fizmatlit, Moscow, 2004 (in Russian).

[364] A.I. Povolotskii, E.A. Gango. Periodic solutions of differential equations with a multi-valued right hand-side. *Leningrad. Gos. Ped. Inst. Uchen. Zap.* 464 (1970), 235–242 (in Russian).

[365] A.I. Povolotskii, E.A. Gango. Periodic solutions of differential equations with a multivalued right hand side. *Leningrad. Gos. Ped. Inst. Uchen. Zap.* 541 (1972), 145–154 (in Russian).

[366] T. Pruszko. Topological degree methods in multi-valued boundary value problems. *Nonlinear Anal.: TMA* 5 (1981), no. 9, 959–970.

[367] B.N. Pshenichnyi. *Convex Analysis and Extremal Problems*. Nauka, Moscow, 1980 (in Russian).

[368] D. Repovš, P.V. Semenov. *Continuous Selections of Multivalued Mappings*. Kluwer, Dordrecht, 1998.

[369] B. Ricceri. Une propriété topologique de l'ensemble des points fixes d'une contraction multivoque á valeurs convexes. *Atti Accad. Naz. Lincei Rend. Cl. Sci. Fis. Mat. Natur.* (8) 81 (1987), no. 3, 283–286 (1988).

[370] B. Ricceri. On the topological dimension of the solution set of a class of nonlinear equations. *C. R. Acad. Sci. Paris Sér. I Math.* 325 (1997), no. 1, 65–70.

[371] F. Riesz, B. Sz.-Nagy. *Functional Analysis.* Dover Publications, Inc., Mineola, NY, 1990.

[372] R.T. Rockafellar. Measurable dependence of convex sets and functions on parameters. *J. Math. Anal. Appl.* 28 (1969), 4–25.

[373] R.T. Rockafellar. Integral functionals, normal integrands and measurable selections. *Lecture Notes in Math.* 543, Springer, Berlin, 1976, 157–207.

[374] V.A. Rokhlin. On the decomposition of a dynamical system into transitive components. *Mat. Sbornik (N.S.)* 25(67), (1949), 235–249 (in Russian).

[375] E. Roxin. Stability in general control systems. *J. Differential Equations.* 1 (1965), no. 2, 115–150.

[376] E. Roxin. On generalized dynamical systems defined by contingent equations. *J. Differential Equations.* 1 (1965), no. 2, 188–205.

[377] E. Roxin. On stability in control systems. *J. Soc. Indust. Appl. Math. Ser. A Control* 3 (1965), 357–372.

[378] I.A. Rus. *Generalized Contractions and Applications.* Cluj University Press, Cluj-Napoca, 2001.

[379] L.E. Rybiński. An application of the continuous selection theorem to the study of the fixed points of multivalued mappings. *J. Math. Anal. Appl.* 153 (1990), no. 2, 391–396.

[380] L.E. Rybiński. *Continuous Selections and Variational Systems.* Wyższa Szkola Inżyner., Instytut Matem., Zielona Góra, 1992.

[381] B.N. Sadovskii. Limit-compact and condensing operators. *Russian Math. Surveys* 27 (1972), 85–155.

[382] J. Saint-Raymond. Points fixes des multiapplications á valeurs convexes. *C. R. Acad. Sci. Paris Sér. I Math.* 298 (1984), no. 4, 71–74.

[383] J. Saint-Raymond. Points fixes des contractions multivoques. *Fixed point theory and applications (Marseille, 1989).* Pitman Res. Notes Math. Ser., 252, Longman Sci. Tech., Harlow, 1991, 359–375.

[384] L. Schwartz. *Analyse Mathematique. Vol.I* Hermann, Paris, 1967.

[385] K.S. Sibirskii, A.S. Shube. *Semidynamical Systems. Topological Theory.* Shtiintsa, Kishinev, 1987 (in Russian).

[386] G.V.Smirnov. *Introduction to the Theory of Differential Inclusions.* Amer. Math. Soc., Providence, R.I., 2002.

[387] E.H. Spanier. *Algebraic Topology.* Corrected reprint. Springer-Verlag, New York-Berlin, 1981.

[388] M. Srebrny. *Measurable Selectors of* <u>PCA</u> *Multifunctions with Applications.* Mem. Amer. Math. Soc. 52 (1984), no. 311.

[389] E. Tarafdar, S.K. Teo. On the existence of solutions of the equation $Lx \in Nx$ and a coincidence degree theory. *J. Austral. Math. Soc. Ser. A* 28 (1979), no. 2, 139–173.

[390] E. Tarafdar, P. Watson, X.-Z. Yuan. Random coincidence degree theory with applications to random differential inclusions. *Comment. Math. Univ. Carolin.* 37 (1996), no. 4, 725–748.

[391] M.J. Todd. *The Computation of Fixed Points and Applications.* Lecture Notes in Economics and Mathematical Systems, 124. Springer-Verlag, Berlin-New York, 1976.

[392] A. Tolstonogov. *Differential Inclusions in a Banach Space.* Mathematics and its Applications, 524. Kluwer Academic Publishers, Dordrecht, 2000.

[393] A.A. Tolstonogov. Solutions of evolution inclusions. I. *Siberian Math. J.* 33 (1992), no. 3, 500–511 (1993).

[394] A.A. Tolstonogov, Ya.I. Umanskii. Solutions of evolution inclusions. II. *Siberian Math. J.* 33 (1992), no. 4, 693–702 (1993).

[395] V. Trenoguine. *Analyse Fonctionnelle.* Mir, Moscow, 1985.

[396] Ya.I. Umanskii. On a property of the solution set of differential inclusions in a Banach space. *Differential Equations.* 28 (1992), no. 8, 1091–1096 (1993).

[397] Ya.I. Umanskii. On the existence of solutions of a class of evolution inclusions on a closed set. I. *Differential Equations.* 30 (1994), no. 7, 1056–1063 (1995).

[398] Ya.I. Umanskii. On the existence of solutions of a class of evolution inclusions on a closed set. II. *Differential Equations.* 30 (1994), no. 12, 2016–2019 (1995).

[399] K. Urai. *Fixed Points and Economic Equilibria.* Series on Mathematical Economics and Game Theory, 5. World Scientific Publishing Co. Pte. Ltd., Hackensack, NJ, 2010.

[400] A. Vanderbauwhede. On a modified degree theory for multivalued mappings. *Simon Stevin.* 50 (1976/77), no. 2, 65–86.

[401] M. Väth. *Topological Analysis. From the Basics to the Triple Degree for Nonlinear Fredholm Inclusions.* De Gruyter Series in Nonlinear Analysis and Applications, 16. Walter de Gruyter, Berlin, 2012.

[402] J. Warga. *Optimal Control of Differential and Functional Equations.* Academic Press, New York-London, 1972.

[403] T. Ważewski. Systèmes de commande et équations au contingent. *Bull. Acad. Polon. Sci. Sér. Sci. Math. Astronom. Phys.* 9 (1961), no. 3, 151–155.

[404] T. Ważewski. Sur une condition équivalente à l'équation au contingent. *Bull. Acad. Polon. Sci. Sér. Sci. Math. Astronom. Phys.* 9 (1961), no. 12, 865–867.

[405] J.R.L. Webb. On degree theory for multivalued mappings and applications. *Boll. Un. Mat. Ital.* (4) 9 (1974), 137–158.

[406] R. Węgrzyk. Fixed-point theorems for multivalued functions and their applications to functional equations. *Dissertationes Math. (Rozprawy Mat.)* 201 (1982), 1–28.

[407] S.A. Williams. An index for set-valued maps in infinite-dimensional spaces. *Proc. Amer. Math. Soc.* 31 (1972), 557–563.

[408] K. Yosida. *Functional Analysis.* Sixth edition. Grundlehren der Mathematischen Wissenschaften, 123. Springer-Verlag, Berlin-New York, 1980.

[409] G.X.–Z. Yuan. *KKM Theory and Applications in Nonlinear Analysis.* Marcel Dekker, New York, 1999.

[410] S.K. Zaremba. Sur une extension de la notion d'equation differentielle *C.R. Acad. Sci. Paris.* 199 (1934), n. 10, 545–548.

[411] S.K. Zaremba. Sur les equations au paratingent. *Bull. Sci. Math.* 60 (1936), n. 2, 139–160.

[412] P. Zecca. Soluzioni periodiche per un'equazione d'evoluzione multivoca. *Boll. Un. Mat. Ital. A (5)* 15 (1978), no. 1, 140–146.

[413] P. Zecca, P.L. Zezza, Nonlinear boundary value problems in Banach spaces for multivalue differential equations on a noncompact interval. *Nonlinear Anal.* 3 (1979), no. 3, 347–352.

[414] E. Zeidler. *Nonlinear Functional Analysis and Its Applications. I. Fixed-Point Theorems.* Springer-Verlag, New York, 1986.

[415] M.Z. Zgurovsky, V.S. Mel'nik, P.O. Kasyanov. *Evolution Inclusions and Variation Inequalities for Earth Data Processing I. Operator Inclusions and Variation Inequalities for Earth Data Processing.* Advances in Mechanics and Mathematics, 24. Springer-Verlag, Berlin, 2011.

[416] M.Z. Zgurovsky, V.S. Mel'nik, P.O. Kasyanov. *Evolution Inclusions and Variation Inequalities for Earth Data Processing II. Differential-Operator Inclusions and Evolution Variation Inequalities for Earth Data Processing.* Advances in Mechanics and Mathematics, 25. Springer-Verlag, Berlin, 2011.

[417] M.Z. Zgurovsky, P.O. Kasyanov, O.V. Kapustyan, J. Valero, N.V. Zadoianchuk. *Evolution Inclusions and Variation Inequalities for Earth Data Processing. III. Long-Time Behavior of Evolution Inclusions Solutions in Earth Data Analysis.* Advances in Mechanics and Mathematics, 27. Springer, Heidelberg, 2012.

[418] Y. Zhou. *Fractional Evolution Equations and Inclusions: Analysis and Control.* Elsevier/Academic Press, London, 2016.

[419] V. Zvyagin, V. Obukhovskii, A. Zvyagin. On inclusions with multivalued operators and their applications to some optimization problems. *J. Fixed Point Theory Appl.* 16 (2014), no. 1–2, 27–82.

Index

CPSIA information can be obtained
at www.ICGtesting.com
Printed in the USA
LVHW082013070420
652563LV00003B/7